T0311464

A COURSE ON
STATISTICS
FOR
FINANCE

A COURSE ON STATISTICS FOR FINANCE

Stanley L. Sclove

CRC Press
Taylor & Francis Group
Boca Raton London New York

CRC Press is an imprint of the
Taylor & Francis Group, an **informa** business

A CHAPMAN & HALL BOOK

MATLAB® is a trademark of The MathWorks, Inc. and is used with permission. The MathWorks does not warrant the accuracy of the text or exercises in this book. This book's use or discussion of MAT-LAB® software or related products does not constitute endorsement or sponsorship by The MathWorks of a particular pedagogical approach or particular use of the MATLAB® software.

First published 2013 by CRC Press

Published 2019 by CRC Press
Taylor & Francis Group
6000 Broken Sound Parkway NW, Suite 300
Boca Raton, FL 33487-2742

First issued in paperback 2020

© 2013 by Taylor & Francis Group, LLC
CRC Press is an imprint of the Taylor & Francis Group, an informa business

No claim to original U.S. Government works

ISBN 13: 978-0-367-57660-8 (pbk)
ISBN 13: 978-1-4398-9254-1 (hbk)

Visit the Taylor & Francis Web site at
http://www.taylorandfrancis.com

and the CRC Press Web site at
http://www.crcpress.com

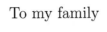

Contents

II   REGRESSION   61

4   Simple Linear Regression; CAPM and Beta   63

List of Figures

List of Tables

Preface

This text has been developed as both a text for university courses and for use by financial analysts and researchers. As a textbook, it is for a second course in statistics, specializing in the direction of financial investments analysis.

Readers wanting a review of basic statistics could read any one of a number of books but one that packs a lot of information into a short space is David Hand's very short introduction (2008). Among basic business statistics books that we have used with success in our department are those by Moore, McCabe, Craig, Alwan, and Duckworth (2011); McClave, Benson, and Sincich (2010); or Levine, Stephan, Krehbiel, and Berenson (2011). These books are listed at the end of this preface. An excellent book that is just above the level of a first course is that by Box, Hunter, and Hunter (2005, first edition 1978).

Little or no background in finance is assumed, although it is believed that even those with some such background might profit from reading the book. Some familiarity with determinants is assumed, such as being able to compute the determinant of two-by-two and three-by-three matrices. Calculus and vectors are used at points in the book, but slowly and carefully. Further, there are appendices relating to some of the more advanced topics.

So, is this a book on "applied" statistics or on "mathematical" statistics? The answer is: both, mixed together. At times there is exposition bordering on a mathematical proof, and at other times there is discussion of how to dump data into software.

It is hoped that beginners come away *both* with improved skills in looking at data *and* with a deeper understanding of the process of modeling. I view this process as perhaps first conceptual, then verbal, and then mathematical.

Main Topics of the Book

The book begins with a review of basic statistics. This includes descriptive statistics (averages, measures of variability, and histograms) and a discussion of types of variables (numerical, non-numerical), derived variables (such as ratios and rates of return), and types of datasets (univariate, multivariate, two-way, seasonal). The book moves relatively soon into *regression analysis*, which is discussed in general terms and also in terms of financial investment models such as the Capital Asset Pricing Model (CAPM) and the Fama/French model. There is an introduction to mean-variance portfolio analysis. Finally, there are chapters relating to time series analysis.

Software

The book is not geared toward any one statistical software package. There will be some mention of Microsoft® Excel™ and of statistical computer packages in general. (My experience has been shaped by varied amounts of use of MINITAB®, SAS®, SPSS®, R®, and MATLAB®).[1] Occasionally, sample output will be shown from MINITAB, slightly edited.

Organization of the Text

Parts of the Book

The parts of the book, consisting of two or three chapters each, are Introductory Concepts and Definitions, Regression, Portfolio Analysis, and Time Series Analysis.

Chapter 1 concerns basic statistics but discusses somewhat more advanced topics because this text is for a second course. Chapter 2 introduces stock price series and rates of return, both ordinary and continuous. Chapter 3 introduces covariance and correlation, and looks in turn at two stocks, three stocks, and m stocks. Because many readers will have had an introduction to regression in an earlier course, Chapter 4, on simple linear regression, pushes this topic a bit further than in a first course. An example in Chapter 4 is the CAPM. Chapter 5 is a discussion of multiple regression, an example being the Fama/French three-factor model, as well as the four-factor model. Chapter 6 discusses bi-criterion portfolio analysis, at the same time introducing some single criteria such as the Sharpe ratio and VaR (Value at Risk). Chapter 7 introduces a single criterion based on a functional derived from expected

---

[1]Microsoft® and Excel™ are trademarks of Microsoft Corporation in the United States, other countries, or both. MINITAB® and all other trademarks and logos for the company's products and services are the exclusive property of Minitab Inc. See minitab.com for more information. SAS® and all other SAS Institute Inc. product or service names are registered trademarks or trademarks of SAS Institute Inc. in the USA and other countries. ® indicates USA registration. R Development Core Team (2008). SPSS® is a registered trademark of IBM Corporation © 2012. All Rights Reserved. R: A language and environment for statistical computing. R Foundation for Statistical Computing, Vienna, Austria. http://www.R-project.org. MATLAB® is © 2012 The MathWorks, Inc. MATLAB is a registered trademark of The MathWorks, Inc.

exponential utility for investor wealth. Chapter 8 is a brief introduction to Box/Jenkins ARIMA models. Chapter 9 considers some definitions of Bull and Bear markets and discusses some ways of segmenting financial time series into such states.

It is possible to cover all the chapters in a semester (averaging a litle less than two weeks per chapter). Sections marked with * are more advanced or not in the mainstream of the development and may be considered optional. To cover all sections in the book or to move at a more leisurely pace, two semesters could be used.

There are several appendices: Appendix A on vectors and matrices; Appendix B on Normal distributions (univariate and multivariate); and Appendix C on Lagrange multipliers. Although notation is defined when introduced, abbreviations and symbols are listed in Appendix D.

Exercises, Mathematical Exercises, Appendices

Exercises appear at the end of some sections and at the end of every chapter. Additionally, at the ends of chapters there are some mathematical exercises. At the end of each chapter there is a list of references. There are appendices in some chapters; these are not side issues and students are advised to read them.

MATLAB® is a registered trademark of The MathWorks, Inc. For product information, please contact: The MathWorks, Inc.
3 Apple Hill Drive
Natick, MA 01760-2098 USA
Tel: 508 647 7000
Fax: 508-647-7001
E-mail: info@mathworks.com
Web: www.mathworks.com

Acknowledgments

I thank the workshop attendees and students who have read earlier forms of this book as class notes. Special thanks are due to Professor Niklas Wagner of Passau University, Germany, who reviewed the manuscript, and to David Grubbs, editor, Joselyn Banks-Kyle, project coordinator for editorial project development, and Karen Simon, project editor, at Chapman and Hall/CRC Press/Taylor & Francis Group. Last, but not least, I thank my family!

Stanley L. Sclove

University of Illinois at Chicago
Chicago, Illinois

"All models are wrong, but some are useful."

—George Box
(1979, section heading, p. 2)

"Statistics is not a discipline like physics, chemistry or biology where we study a subject to solve problems in the same subject. We study statistics with the main aim of solving problems in other disciplines."

—C.R. Rao

"He uses statistics as a drunken man uses lamp posts - - for support rather than for illumination."

—Andrew Lang
(1844–1912), Scottish poet

Bibliography

Berenson, Mark L., Levine, David M., and Krehbiel, Timothy C. (2012). *Basic Business Statistics, 12th ed.* Pearson (Prentice Hall), Upper Saddle River, NJ.

Box, George E. P. (1979). Robustness in the strategy of scientific model building. *Robustness in Statistics: Proceedings of a Workshop (at Army Research Office)*, R. L. Launer and G. N. Wilkinson (Eds.). Academic Press, New York.

Box, George E. P., Hunter, William G., and Hunter, J. Stuart (2005). *Statistics for Experimenters: An Introduction to Design, Data Analysis, and Model Building. 2nd ed.* John Wiley & Sons, Inc., New York. (First edition, 1978.)

Hand, David J. (2008). *Statistics: A Very Short Introduction.* Oxford University Press, Oxford, UK; New York, NY.

Levine, David M., Krehbiel, Timothy C., and Berenson, Mark L. (2010). *Business Statistics: A First Course. 5th ed.* Pearson (Prentice Hall), Upper Saddle River, NJ.

Levine, David M., Stephan, David F., Krehbiel, Timothy C., and Berenson, Mark L. (2011). *Statistics for Managers Using Microsoft Excel, 6th ed.* Pearson (Prentice Hall), Upper Saddle River, NJ.

McClave, James T., Benson, P. George, and Sincich, Terry (2010). *Statistics for Business and Economics. 11th ed.* Pearson (Prentice Hall), Upper Saddle River, NJ.

Moore, David S., McCabe, George P., Craig, Bruce, Alwan, Layth, and Duckworth, Wm., III. (2011). *The Practice of Statistics for Business and Economics. 3rd ed.* W. H. Freeman Co., New York.

About the Author

Stanley L. Sclove (A.B., applied honor mathematics, Dartmouth College; Ph.D., mathematical statistics, Columbia University) is a professor of statistics in the Department of Information and Decision Sciences of the College of Business Administration at the University of Illinois at Chicago (UIC). In addition to UIC he has taught at Carnegie Mellon, Northwestern, and Stanford universities. Sclove's areas of specialization within statistics include multivariate statistical analysis, cluster analysis, time series analysis, and model selection criteria. He has taught courses in a number of areas of mathematics, probability, and statistics, including especially applied statistical methods, regression analysis, time series analysis, multivariate statistical analysis, and structural equation modeling. Sclove's research interests include time series segmentation and regime switching via Markov models.

Sclove is author or co-author of articles in a number of statistical and scientific journals and co-author of several books on statistical data analysis and business statistics. He has directed a number of doctoral dissertations. He is a frequent referee and reviewer. Sclove is a member of a number of professional societies and an officer of the Classification Society and the Section of Risk Analysis of the American Statistical Association.

Part I

INTRODUCTORY CONCEPTS AND DEFINITIONS

1

Review of Basic Statistics

CONTENTS

1.1   What Is Statistics?

This chapter is a review of basic statistics. It begins with a discussion of the nature of data, variables, and statistical analysis. Then, in view of the fact that this book is mainly for a second course on statistics, the chapter proceeds with a few nonelementary items.

1.1.1 Data Are Observations

Data result from the observation of one or more variables. In the context of statistics, a *variable* represents a characteristic or property that can be observed or measured. Variables may may be observed on a number of occasions, or for a number of individuals (or for a number of individuals on a number of occasions).

1.1.2 Statistics *Are* Descriptions; Statistics *Is* Methods

Statistics (*plural*) are numerical *descriptions* of data, such as percentages and averages.

Statistics (*singular*) is the body of *methods* used to deal with data, by computing and interpreting Statistics (plural) and thus transforming data into information. *Information* is data summarized and conceptualized. Information forms a basis for decisions.

1.1.3 Origins of Data

The word *data* is the plural past participle of the Latin word "to give," so "data" are "givens."

Data are obtained within a particular situation. They may concern individual people, groups of people, or objects. *Financial data* include observations of such variables as prices of stocks and levels of stock indices.

1.1.4 Philosophy of Data and Information

TABLE 1.1
Data to Information to Decision to Action

	Statistical Analysis		Decision Analysis		Management	
DATA	——>	INFORMATION	——>	DECISION	——>	ACTION

1.1.4.1 Data versus Information

Most people seem to believe that correct information, gleaned from data, leads somehow to the truth.

Truth and Information

There is a Russian saying that contrasts truth and information. The word *izvestya* means *information*. The word *pravda* means *truth*. These two words were the names of the major newspapers in Russia. (*Pravda* was the official newspaper of the Central Committee of the Communist Party between 1912 and 1991. *Izvetya* was the official newspaper of the Soviet government.

About these newspapers it was said: "In *Izvestya*, no truth; in *Pravda*, no information." (In "Information," no truth; in "Truth," no information.)

A dataset can hide the real information it contains. This is perhaps particularly true of large datasets. Underlying patterns must be found to reveal the essence of what is there. This is one of the tasks of Statistics. *Statistical Analysis* transforms *Data* into *Information*.

"Uncertainty . . .

   Something you can always count on."

slogan on T-shirt
–American Statistical Association

Variability is inherent in the processes of observation and measurement. Managers and financial analysts need to use statistical analysis because variation is everywhere, important patterns may not be obvious, and conclusions are not certain. Decisions are thus made in an atmosphere of risk.

1.1.4.2   Decisions

Decisions are based on prior experience, expert opinion, and information gleaned from data. Decisions consider costs and benefits. *Decision Analysis* (also called *Decision Risk Analysis*) transforms Information into Decisions. A diagrammatic tool that is used in this sort of analysis is the *decision tree*. The branches represent different alternative decisions, which are labeled with their probabilities, costs, and profits or other benefits. Some universities have courses on decision risk analysis; sometimes the topic is included in courses on operations research, operations management, or management science. Some textbooks on decision risk analysis (Clemen and Reilly 2004, Golub 1997) are listed in the Bibliography.

The diagram (Table 1.1) shows the progression from Data to Information to Decisions to Action. The purpose of Statistical Analysis is the transformation of Data into Information. This transformation is accomplished by means of Statistical Analysis. Decision Risk Analysis weighs costs against benefits and forms a basis for making decisions based on information. This book is concerned mostly with the Statistical Analysis portion of this diagram. As a beginning, ways of describing and summarizing data will be discussed.

1.2   Characterizing Data

As stated above, in the context of statistics, a *variable* represents a characteristic or property that can be observed or measured. Variables will be denoted by symbols such as X and Y or by more specific symbols such as h for height or P for price. The values of a variable X for a sample of n individuals will be denoted by $x_1, x_2, \ldots, x_n$. Usually the discussion centers on a sample rather than a population. To make a distinction, the values for a *population* of N individuals could be denoted by $\xi_1, \xi_2, \ldots, \xi_N$. This is in keeping with the custom of denoting sample quantitites by Latin letters and the corresponding population quantitites by the corresponding Greek letters.

1.2.1   Types of Data

Perhaps the most common type of dataset is a rectangular array of cases by variables. Such would be the case for a roster of students, with the major and year for each. The *cases* are individual persons or firms. Think of them as the rows (or *records*) in a spreadsheet. The *variables* are properties or characteristics of the cases. Think of them as the columns (or *fields*) in a spreadsheet.

1.2.1.1   Modes and Ways

More generally, data can be characterized in terms of *modes, ways,* and *levels.* (See esp. Carroll and Arabie 1980). An array of cases by variables is an example of two-mode, two-way data. It is two-way because it is two-dimensional, with rows and columns. It is two-mode, the modes being cases and variables.

An example of *one-mode, two-way data* is a mileage chart, with the names of cities down the side and across the top, and the entries of the table being the distances between the cities.

There is *three-way, three-mode data*. This can be thought of as a *data cube*. A cube has length, width, and height. A data cube can be considered in terms of such dimensions, with subjects, variables, and occasions of measurement along the axes.

1.2.1.2   Types of Variables

Variables can be non-numerical or measured on a numerical scale. Stevens (1966) classified *levels of measurement* as nominal, ordinal, interval-scale, or ratio-scale. Non-numerical variables can be nominal or ordinal. Nominal variables include such things as names or eye color. Ordinal variables can be recorded as low versus high, or low, medium, or high. *Likert scale* items are those where, given a statement, the subject indicates strong agreement, agreement, neutrality, disagreement, or strong disagreement. This is a five-point Likert scale, perhaps the most frequently used one. Seven-point and four-point scales are also common. Ratio-scale variables, such as height, weight, price, and quantity, have have a meaningful zero. Interval-scale variables, such as Fahrenheit or Celsius temperature, are numerical but the zero may not have special meaning: it does not signify the absence of heat. (On the absolute, or Kelvin, temperature scale, zero means the absence of heat in the sense of the absence of molecular motion.) Likert scales are often treated as interval scales, although they really are not. This may or may not make a big difference.

Some numerical variables exhibit a bell-shaped Normal distribution. (That is, the distribution is shaped like the cross-section of a bell, high in the middle with the frequency falling off in either direction.)

1.2.1.3   Cross-Sectional Data versus Time Series Data

A single *time series* consists of a single variable recorded over time. This is one-way, one-mode data, indexed by time t.

For stock prices P_t people consider daily, weekly, monthly, or annual prices, that is, time t could be in days, weeks, month, quarters, or years. The data could also be recorded for each transaction ("tick by tick").

Multiple time series consist of several single time series. Consider the prices P_{it} of stocks $i = 1, 2, \ldots, m$, at times $t = 1, 2, \ldots, n$. For each fixed stock i, the prices P_{it}, $t = 1, 2, \ldots, n$, constitute a time series. Alternatively, the series may be considered in terms of vectors $\boldsymbol{p}_t$, $t = 1, 2, \ldots, n$, where $\boldsymbol{p}_t$ is the vector $(P_{1t}\, P_{2t}, \ldots, P_{mt})'$. For a fixed time t, the set of prices $P_{it}, i = 1, 2, \ldots, m$, is cross-sectional data.

1.2.2   Raw Data versus Derived Data

Sometimes two or more variables are processed into a single new variable before analysis. For example, physical work is the result of a multiplication, the product of a distance and a weight. Units of work are newton-meters (joules). Physical force is the result of a multiplication, the product of mass and acceleration. *Ratios* are of course the result of division.

1.2.2.1 Ratios

A ratio is the result of dividing one number, a numerator (or dividend), by another, a denominator (or divisor). The resulting quotient is a ratio. Thus, ratios are derived data, but they may be analyzed on their own. Examplesof ratios are fuel efficiency, fuel consumption, body-mass index, and financial rates of return.

Given runs of a car, $i = 1, 2, \ldots, n$, and the distance traveled (in kilometers), d_i, and liters of gasoline g_i used in the ith run, the kilometers per liter for the ith run is the ratio d_i/g_i. If d_i is in miles and g_i is in gallons, the ratio is in miles per gallon (MPG).

The fuel efficiency ratio is derived data, but it may be analyzed as a dependent variable, as a function of various conditions, such as the type of road and the type of fuel used. The measure kilometers per liter is usually abbreviated as km/L. The reciprocal ratio, fuel consumption, would be expressed in liters per 100 kilometers (L/100 km) or gallons per mile.

1.2.2.2 Indices

An *index* is another example of derived data. Given $i = 1, 2, \ldots, n$ persons, and their heights h_i and weights w_i, the *body-mass index* (BMI) is w_i/h_i^2, where the height is in meters and the weight in kilograms. The BMIs are then data derived from the heights and weights. As an example, if a man weighs 80 kg and is 1.76 m tall, his BMI is $80/1.76^2 = 25.8$. (To convert to English units, write $\text{kg/m}^2 = (\text{lbs.}/2.2046)/(2.54\text{in.}/100)^2 = 703.1 \text{ lb/in}^2$.) BMI, being computed from height and weight, is a derived variable, but may be analyzed as if it were raw data, perhaps as a function of various health and nutrition factors. (BMI was invented by Adolphe Quetelet—a Belgian polymath, in his case, astronomer, mathematician, statistician and sociologist—between the years 1830 and 1850.)

An economic index is the *consumer price index* (CPI). It is the cost at any fixed point in time of a standard market basket of goods. Stock market indices are weighted averages of prices of specified sets of stocks, where the weights may be, for example, the capitalizations of the companies.

Specific financial variables that are derived variables, such as rates of return, will be introduced in the next chapter and revisited in later chapters on portfolio analysis.

1.3 Measures of Central Tendency

This section is concerned with measuring the location or center of sets of data. Measures of central tendency include the mode, median, and mean. Many of the concepts apply both to populations and samples, but usually here the notation and discussion will be in terms of samples.

1.3.1 Mode

The *mode* is one measure of the location of a set of observations. The mode is the most frequently occurring value. To take a non-numerical example, if the variable is first name, and its values in a sample are Jim, Jeff, Stan, Mike, Judy, Jim, Norm, Dave, Bill, Mark, Gary, Jeff, Jim, Betty, Jerry, Randy, and Rudy, then there are two Jeffs, three Jims, and one each of Stan, Mike, Judy, Norm, Dave, Bill, Mark, Gary, Jerry, Randy, Betty, and Rudy, so the name Jim is the mode, because the name Jim occurs more often than any other single name. However, this mode is not particularly outstanding, as Jeff is a close second, with two, and the distribution is flat anyway, with 14 names for 17 people. The variable here is non-numerical. For a numerical variable, the mode can be more meaningful when the frequencies of values near it are also relatively high. Also, modes have more meaning with the distributions of two or more groups. Consider, for example, adult male and female heights to the nearest centimeter. The mode for males might be 178 cm. while that for females might be 165 cm., 13 cm. lower. The modes are descriptive in this case because presumably nearby values would also be frequent.

1.3.2 Measuring the Center of a Set of Numbers

An indication of the location or center of a set of numbers is often called the *average*. The word "average" comes from a root referring to loss or damage in maritime shipping (*Oxford English Dictionary*). The word came to refer to measuring such loss in financial terms. The parties involved would agree to be equally (or proportionally) responsible for such loss. The word "average" came to refer to each party's share.

Suppose that a number a is considered as a candidate for the "average" of a set of numbers. Then the chosen value a should be in the center of the set, in some sense.

1.3.2.1 Median

The *median* is one measure of the center of a set of numbers. Suppose the heights of a set of 7 men are 170, 181, 176, 175, 177, 182, 165 cm. Put in order, these are 165, 170, 175, 176, 177, 181, 182. This ordered list is the *order statistic* of the sample. The median is the height of the man in the center. That is the fourth ranking height, and it is 176 cm. The mean is 176 cm. The median is the middle number. Let n denote the number of individuals in the sample. That is, if n is odd, the median is located $(n+1)/2$ observations from the beginning of the ordered list.

There are various definitions of the median. Generally, if n is odd, say $n = 2m + 1$, then the m-th ranking value is the median. If n is even, say $n = 2m$, then the median can be taken as the number half-way between the m-th and the $(m + 1)$-st. However, it is usually preferable to group the data into consecutive categories ("bins") and estimate a median by interpolation on the bin frequencies to reach a cumulative relative frequency of one-half.

1.3.2.2 Quartiles

The *quartiles* divide a set of numbers into quarters. They are the first (lower) quartile, the second quartile (the median), and the third (upper) quartile. The *order statistic* of a sample $x_1, x_2, \ldots, x_n$ means the sample sorted in ascending order. It is often denoted by $x_{(1)}, x_{(2)}, \ldots, x_{(n)}$.

If $n = 4m + 1$, the lower quartile Q_1 is $x_{(m)}$ and the upper quartile Q_3 is $x_{(3m)}$. However, as remarked in the case of the median, it is often better to group the data into bins and estimate the quartiles by interpolation on the bin frequencies to reach cumulative relative frequencies of one-fourth and three-fourths.

As far as terminology is concerned, it is perhaps better to say "lower" and "upper" quartile than first and third quartile, because the use of the words "first" and "third" assumes that you know you are working from low to high. The lower and upper quartiles can be defined as the medians of the lower and upper halves of the sample. A *five-number summary* is useful for indicating location: the *minimum, lower quartile, median, upper quartile,* and *maximum. Box plots* show the quartiles and the min and max. The median is also added to the plot.

1.3.2.3 Percentiles

The $100p$-th *percentile* of the distribution of a random variable x is the value x_p which is exceeded with probability $1 - p$. Percentiles and quartiles are of course defined both for distributions and for datasets. The lower quartile is $x_{.25}$; the upper quartile, $x_{.75}$. The second quartile $x_{.5}$ is the median.

For a standard Normal variable Z, the 95-th percentile is $z_{.95} = 1.645$ and the fifth percentile is $z_{.05} = -1.645$. Percentiles of Z can be obtained from tables, in spreadsheet software, or in statistical software.

A general term which includes both *quartile* and *percentile* is *quantile*.

1.3.2.4 Section Exercises

1.1 Given $n = 8$ observations 170, 190, 173, 174, 176, 177, 175, 179, find the median.

1.2 Given $n = 8$ observations 170, 190, 173, 174, 176, 176, 175, 179, find the median. *Hint:* Same answer as preceding exercise.

1.3 The numbers of children in $n = 10$ families are 3, 0, 0, 1, 1, 4, 2, 5, 2, 2. Find the median.

1.4 (continuation) Make a table of two columns, the values 0, 1, 2, 3, 4, 5, of the numbers of children, and their frequencies.

1.5 (continuation) Make a histogram, that is, a bar graph with the values along the horizontal axis and the frequencies along the vertical axis.

1.6 (continuation) Estimate the probability of a family's having two children. This probability is the proportion of the population of families having exactly two children.

1.3.2.5   Mean

Next, another way of defining a number, say a, as an "average" of a set of numbers will be derived. Such a number a to be in the center of the set, in some sense. The deviations from a are $x_1 - a$, $x_2 - a$, ..., $x_n - a$. When $x_i > a$, the deviation $x_i - a > 0$. When $x_i < a$, the deviation $x_i - a < 0$. The sum of deviations is $(x_1 - a) + (x_2 - a) + \cdots + (x_n - a)$. When a is in the center of the dataset, this sum should be zero.

Given values

$$x_1, x_2, \ldots, x_n,$$

the sum is denoted using *summation notation* by

$$\sum_{i=1}^{n} x_i = x_1 + x_2 + \cdots + x_n.$$

The sum of deviations from a is denoted by $\sum_{i=1}^{n} (x_i - a)$. What kind of average is in the center of the set of values in the sense that $\sum_{i=1}^{n} (x_i - a) = 0$? To answer this question, note that $\sum_{i=1}^{n} (x_i - a) = \sum_{i=1}^{n} x_i - \sum_{i=1}^{n} a = \sum_{i=1}^{n} x_i - na = 0$, Solving this for a gives $a = \sum_{i=1}^{n} x_i/n$, often denoted by $\bar{x}$. This is called the *mean* or *ordinary arithmetic average;* it is the sum over (divided by) the number.

Deviations from the mean.   The *deviations* from the mean are

$$x_1 - \bar{x}, x_2 - \bar{x}, \ldots, x_n - \bar{x}.$$

The sum of the deviations about the mean is zero; this is a sense in which the mean is at the center of the sample. To see this, write

$$\sum_{i=1}^{n} (x_i - \bar{x}) = \sum_{i=1}^{n} x_i - \sum_{i=1}^{n} \bar{x} = \sum_{i=1}^{n} x_i - n\bar{x} = n\bar{x} - n\bar{x} = 0.$$

The number of *degrees of freedom* is the number of independent quantities. There are n deviations from the mean, but only $n - 1$ independent quantitites among them, hence these n quantities have $n - 1$ degrees of freedom. If you tell me the value of the sum of $n - 1$ of the deviations, I can tell you the value of the n-th, because the sum of all n deviations is zero. For example, if you tell me that the first $n - 1$ deviations sum to -3, then I know that the n-th deviation is $+3$. As discussed in an earlier section, the mean is at the center of the sample, in that the size of the sum of the negative deviations equals the sum of the positive deviations. For example, given heights 170, 171, 173, 175, 177, 177, 178, 180, 183 cm, the mean is 176 cm, and the deviations from the mean are $-6, -5, -3, -1, 1, 1, 2, 4$, and 7 cm. The size of the sum of the negative deviations is $6 + 5 + 3 + 1 = 15$, and so is the sum of the positive deviations, $1 + 1 + 2 + 4 + 7 = 15$.

A note on pronunciation. Above, the phrase "arithmetic average" was used. Note that the four-syllable word *arithmetic* as an adjective is pronounced a-rith-*me*-tic, with the accent on the third syllable. As a noun, it is pronounced a-*rith*-me-tic, with the accent on the second syllable.

1.3.2.6 Other Properties of the Ordinary Arithmetic Average

What kind of average, a, represents all the number in the set in the sense that it will give the total when multiplied by the number n of cases? To answer this, solve $n\,a = \sum_{i=1}^{n} x_i$. This gives $a = \sum_{i=1}^{n} x_i/n = \bar{x}$. As an example, if the mean weight of 200 airline passengers is 80 kg, then the total weight is $200 \times 80\text{kg.} = 16,000$ kg.

There is another way in which the mean $\bar{x}$ represents the center of the whole set. Suppose that the set of observations $(x_1\, x_2\, \ldots\, x_n)$ is to be replaced by $(a\,a\ldots a)$. What choice of a gets closest to $(x_1\, x_2\, \ldots\, x_n)$, in the sense of minimizing the distance between the two points $(a\, a\, \ldots\, a)$ and $(x_1\, x_2\, \ldots\, x_n)$? Now, in general, the geometric distance between two points $(x_1\, x_2\, \ldots x_n)$ and $(y_1\, y_2\, \ldots\, y_n)$ is

$$\sqrt{\sum_{i=1}^{n} (x_i - y_i)^2}.$$

Take $(y_1\, y_2 \ldots y_n)$ to be $(a\,a\ldots a)$. Then it is seen that the distance between $(x_1\, x_2\, \ldots\, x_n)$ and $(a\, a\, \ldots\, a)$ is

$$\sqrt{\sum_{i=1}^{n} (x_i - a)^2}.$$

We have

$$\min_a \sqrt{\sum_{i=1}^n (x_i - a)^2} = \sqrt{\min_a \sum_{i=1}^n (x_i - a)^2}$$

$$= \sqrt{\min_a [\sum_{i=1}^n (x_i - \bar{x})^2 + n(\bar{x} - a)^2]}$$

$$= \sqrt{\sum_{i=1}^n (x_i - \bar{x})^2}.$$

Here we have used the fact that $\sum_{i=1}^n (x_i - a)^2 = \sum_{i=1}^n (x_i - \bar{x})^2 + n(\bar{x} - a)^2$. To see this, note that

$$\sum_{i=1}^n (x_i - a)^2 = \sum [(x_i - \bar{x}) + (\bar{x} - a)]^2$$

$$= \sum (x_i - \bar{x})^2 + 2\sum (x_i - \bar{x})(\bar{x} - a) + n(\bar{x} - a)^2$$

$$= \sum (x_i - \bar{x})^2 + n(\bar{x} - a)^2,$$

because $\sum (x_i - \bar{x})(\bar{x} - a) = (\bar{x} - a)\sum (x_i - \bar{x}) = 0$.

The fact that $\bar{x}$ is the value of a which minimizes $\sum_{i=1}^n (x_i - a)^2$ follows, because

$$\sum_{i=1}^n (x_i - \bar{x})^2 + n(\bar{x} - a)^2 \geq \sum_{i=1}^n (x_i - \bar{x})^2,$$

this minimum being achieved when $a = \bar{x}$. The fact that $\bar{x}$ is the minimizing value of a can be derived in a couple of other ways that may be useful later. One of these ways is to expand the square, which means expressing a quadratic as a square of a binomial term, plus other terms. This gives

$$\sum_{i=1}^n (x_i - a)^2 = \sum_{i=1}^n (x_i^2 - 2ax_i + a^2)$$

$$= \sum_{i=1}^n x_i^2 - 2a\sum_{i=1}^n x_i + na^2$$

$$= Ax^2 + Bx + C,$$

with $x = a$, $A = n$, $B = -2\sum_{i=1}^n x_i$, $C = \sum_{i=1}^n x_i^2$. Now, a quadratic $Ax^2 + Bx + C$ obtains its minimum if $A > 0$ or maximum if $A < 0$ at

$x = -B/2A$; this gives $a = -(-2\sum_{i=1}^{n} x_i)/2n = \sum_{i=1}^{n} x_i/n = \bar{x}$. Another way of showing that $\bar{x}$ is the minimizing value of a is to define

$$f(a) = \sum_{i=1}^{n} (x_i - a)^2,$$

differentiate with respect to a, set the derivative equal to zero, and solve for the number a. This gives

$$\begin{aligned}
f'(a) &= \frac{d}{da} \sum_{i=1}^{n} (x_i - a)^2 \\
&= \sum_{i=1}^{n} \frac{d}{da}(x_i - a)^2 \\
&= \sum_{i=1}^{n} (-2)(x_i - a) \\
&= (-2) \sum_{i=1}^{n} x_i + 2na = 0,
\end{aligned}$$

giving $\sum_{i=1}^{n} x_i - na = 0$, or $a = \sum_{i=1}^{n} x_i/n$.

To summarize: Several properties of the sample mean $\bar{x}$ have been discussed:

- $\bar{x}$ is in the center of the set of numbers, in the sense that it is the choice of average a such that $\sum_{i=1}^{n} (x_i - a) = 0$.

- $\bar{x}$ is the choice of average a such that multiplying it by n gives the total: $na = \sum_{i=1}^{n} x_i$.

- $\bar{x}$ is the choice of average a that comes closest to the set of numbers in the sense that it minimizes $\sum_{i=1}^{n} (x_i - a)^2$.

1.3.2.7 Mean of a Distribution

The mean (expected value, mathematical expectation) of a random variable x is denoted by $\mathcal{E}[x]$ or μ_x. If the r.v. X is discrete with values v_j, $j = 1, 2, \ldots, m$, and probability mass function

$$p_X(v_j) = \Pr\{X = v_j\}, \ j = 1, 2, \ldots, m,$$

then

$$\mu_x = \sum_{j=1}^{m} v_j \, p_X(v_j),$$

the probability-weighted average of the possible values of X. If X is continuous with probability density function $f_X(v)$, then

$$\mu_x = \int v \, f_X(v) \, dv.$$

1.3.3   Other Kinds of Averages

There are many kinds of averages.

Example 1.1 Baseball averages

In baseball there are defensive averages like the Earned Run Average (ERA), the number of earned runs a pitcher has allowed, per game equivalent, where the game equivalent is the number of innings pitched, divided by 9.

There are offensive averages like the Batting Average (BA), the number of hits over the number of at bats, and the Slugging Average (abbreviated as SA or SLG). The SA is the total bases per at bat, that is, the mean number of bases per at bat (AB). If in a season a player had 500 ABs, with 80 singles, 50 doubles, 10 triples, and 40 homeruns, then his total bases is $1(80) + 2(50) + 3(10) + 4(40) = 80 + 100 + 30 + 160 = 370$ bases. His SA is (total bases)/AB $= 370/500 = 0.740$.

As of this writing, the record one-season SA is 0.863 by Barry Bonds in 2001, and the record career SA is 0.690 by Babe Ruth (*Baseball Almanac*).

1.3.3.1   Root Mean Square

Given positive numbers, say $a_1, a_2, \ldots, a_n$, their *root mean square* is the square root of the mean of their squares,

$$\text{RMS} = \sqrt{\sum a_i^2/n}.$$

The root mean square of the distances from the mean is a measure of variability. (See the section on measuring variability; Section 1.4.)

1.3.3.2   Other Averages

The root mean square is one of a type of average in which the observations are transformed, the results are averaged, and then the inverse transformation is applied. Given a one-to-one function $h()$, such an average is

$$\text{Average} = h^{-1}\left[\sum_{i=1}^{n} h(x_i)/n\right].$$

Here the notation $h^{-1}(y)$ denotes the inverse of the function $y = h(x)$. An example is $h(x) = 1/x \, (x \neq 0)$. Then the average is $1/\sum_{i=1}^{n}(1/x_i)/n$. This is known as the *harmonic mean*. For positive data, the function $h(x) = \ln x$ can be used. This gives

$$\text{Average} = \exp[\sum_{i=1}^{n} \ln(x_i)/n].$$

This is equal to the n-th root of the product:

$$\exp[\sum_{i=1}^{n} \ln(x_i)/n] = \exp[\ln(\prod_{i=1}^{n} x_i)/n)]$$

$$= \exp\{\ln[(\prod_{i=1}^{n} x_i)^{1/n}]\}$$

$$= [\prod_{i=1}^{n} x_i]^{1//n}.$$

This kind of average is known as the *geometric mean*. Here the natural log has been used; common logs or logs with respect to other bases could be used.

1.3.4   Section Exercises

1.7 Five-number summary. Suppose the heights of 9 men are 170, 166, 181, 176, 175, 177, 182, 179, 165 cm.

   a. Put these values in order.

   b. Find the minimum.

   c. Find the maximum.

   d. Compute the range.

   e. What is the five-number summary?

1.8 Five-number summary. Suppose the heights of 17 men are 172, 170, 173, 166, 181, 176, 175, 177, 178, 184, 182, 167, 180, 183, 179, 165, 174 cm.

   a. Put these values in order.

   b. Find the minimum.

   c. Find the maximum.

   d. Compute the range.

   e. What is the five-number summary?

1.9 Using a working mean. (continuation) Compute the mean of these values, as follows.

   a. Put these values in order.

   b. Subtract 160 from each height.

   c. Compute the mean of these differences from 160.

   d. Add 160 to this result to obtain the mean of the original values. The preliminary number (here 160) is sometimes called a "working mean," although in this caseit is less than all the values. The point is to subtract a number to obtain smaller numbers to average and then add the number back in.

1.10 Baseball slugging average. Show that the slugging average can be written as

$$\text{SA} = (\text{ hits} + \text{ 1 D} + \text{ 2 T} + \text{ 3 H})/\text{AB},$$

where D = no. of doubles, T = no. of triples, and H = no. of homeruns.

1.11 A time series. The daily average temperature (average temperature within a day) is the low plus the high, divided by 2. For Chicago for November 1 to 9, 2008, these were 54.4, 53.8, 48.3, 47.2, 41.3, 48.8, 54.7, 55.2, 52.6 degrees Fahrenheit. The mean of these is 50.7. The deviations from the mean are $+3.7, +3.1, -2.4, -3.5, -9.4, -1.9, +4.0, +4.5, +1.9$. Plot the series of deviations against $1, 2, \ldots, 9$.

1.12 (continuation) There are two positive deviations, followed by four negative deviations, and then three positive deviations. Does this seem to be a random pattern of signs, or is it more suggestive of a correlated time series? Would you expect to toss two heads, then four tails, then three heads, or would you expect the heads and tails to be more mixed?

1.4   Measures of Variability

This section, on *variability,* is concerned with measuring the spread of sets of numbers and distributions. Measures of variability, including the range, interquartile range, IQR;   mean absolute deviation, MAD;   variance, and standard deviation, are discussed.

1.4.1   Measuring Spread

There are positional and distance-based measures of spread.

1.4.1.1 Positional Measures of Spread

Positional measures of spread are based on the space between positional summary statistics; they include the *range* and the *inter-quartile range*.

1.4.1.2 Range

The *range* is the distance between the minimum and maximum; it is the difference, maximum minus minimum. *Box plots* show both center and spread by showing the quartiles and the min and max.

1.4.1.3 IQR

The *inter-quartile range* (IQR) is the difference between the lower and upper quartiles. Half the distribution, in the middle of it, is within this range. In the standard Normal distribution, twenty-five percent of the distribution is less than 0.6745 standard deviations below the mean, that is, the lower quartile is at $z = -0.6745$. The upper quartile is at $+0.6745$. The IQR is $2(0.6745) = 1.349$ standard deviations wide.

1.4.2 Distance-Based Measures of Spread

Distance-based measures of spread are based on an average distance of the cases to their center.

1.4.2.1 Deviations from the Mean

The deviations from the mean are $x_1 - \bar{x}, x_2 - \bar{x}, \ldots, x_n - \bar{x}$. Their sum is zero. The sum of the negative deviations equals the sum of the positive deviations. This is a way in which the mean is in the center of the data.

1.4.2.2 Mean Absolute Deviation

The *mean absolute deviation* (MAD) is the ordinary arithmetic average of the distances to the mean. The absolute values of the deviations are the distances to the mean,

$$|x_1 - \bar{x}|, |x_2 - \bar{x}|, \ldots, |x_n - \bar{x}|.$$

The MAD is

$$\frac{|x_1 - \bar{x}| + |x_2 - \bar{x}| + \ldots + |x_n - \bar{x}|}{n}.$$

That is,

$$\text{MAD} = \sum_{i=1}^{n} |x_i - \bar{x}|/n.$$

If d_i denotes $x_i - \bar{x}$, then the MAD is simply the ordinary arithmetic average of $|d_1|, |d_2|, \ldots, |d_n|$.

1.4.2.3   Root Mean Square Deviation

The *root mean square deviation* is another kind of average distance to the mean. It is the square root of the ordinary arithmetic average of the squared distances to the mean. It is computed by finding the sum of squared deviations from the mean (SSD), dividing it by n, and taking the square root of the result.

1.4.2.4   Standard Deviation

The *standard deviation* of a sample is computed by finding the sum of squared deviations from the mean (SSD), dividing it by $n-1$ (this result is called the *variance*) and taking the square root of the result. So the standard deviation is

$$s = \sqrt{\text{variance}} = \sqrt{\text{SSD}/(n-1)} = \sqrt{\text{SSD}}/\sqrt{n-1}.$$

As discussed above, the ordinary Euclidean distance (ruler distance) between two points $(p_1\, p_2 \,\ldots\, p_n)$ and $(q_1\, q_2\, \ldots\, q_n)$ is

$$\sqrt{(p_1 - q_1)^2 + (p_2 - q_2)^2 + \cdots + (p_n - q_n)^2}.$$

So the numerator of s, namely $\sqrt{\text{SSD}}$, is the ordinary (ruler) distance between the points $(x_1, x_2, \ldots, x_n)$ and $(\bar{x}, \bar{x}, \ldots, \bar{x})$. So measures of variability involving $\sqrt{\text{SSD}}$ are in this way the natural measures of variability. Note again that

$$s = \sqrt{\frac{\text{SSD}}{(n-1)}}.$$

A reason for using a divisor of $n-1$ rather than n is that the n deviations $d_1, d_2, \ldots, d_n$ are not mathematically independent; rather, their sum is 0. That is, if $n-1$ of their values are known, the n-th is determined. It is said that these n quantities satisfy one *constraint*, namely, that their sum be zero, and hence that they have $n-1$ *degrees of freedom*.

So, what *is* the "standard" deviation? Judging from the phrase, it should be some kind of average size of deviation, and it is. More precisely, the sample standard deviation s is related to the distance between the data vector and the vector whose every element is the mean. It is this distance, say D, divided by the square root of $n-1$, that is,

$$s = D/\sqrt{n-1},$$

where

$$D = \sqrt{\sum_{i=1}^{n} (x_i - \bar{x})^2}.$$

That is,

$$s = \sqrt{v},$$

where the sample variance

$$v = \sum_{i=1}^{n} (x_i - \bar{x})^2 / (n - 1).$$

Because of the importance of the standard deviation s, the statistic v is often written as s^2.

Computational formulas. The sum of squared deviations SSD $= \sum_{i=1}^{n} (x_i - \bar{x})^2$ can be computed in several different ways:

$$\text{SSD} = \sum_{i=1}^{n} x_i^2 - n\bar{x}^2 = \sum_{i=1}^{n} x_i^2 - (\sum_{i=1}^{n} x_i)^2 / n.$$

1.4.2.5 Variance of a Distribution

The variance of a random variable X is denoted by $\mathcal{V}[x]$ or σ_x^2. If X is discrete, then

$$\sigma_x^2 = \sum_{j=1}^{m} (v_j - \mu_x)^2 p_x(v_j).$$

If X is continuous, then

$$\sigma_x^2 = \int (v - \mu_x)^2 f_X(v) \, dv.$$

The standard deviation of X is the square root of its variance. It is a particular kind of average distance to the mean, namely, the square root of the average squared distance to the mean.

In a Normal distribution, about two-thirds of the cases (actually about 68%) are within one standard deviation of the mean, about 95% are within two standard deviations of the mean, and about 99.7% are within three standard deviations of the mean. This is called the "68, 95, 99.7 percent" rule for Normal distributions. Of course, in a sample of 20 you would not expect to get any observations more than three standard deviations from the mean, but in a sample of 1,000 you would expect to get several. If adult male height is Normally distributed with a mean of 173 cm and a standard deviation of 7 cm, then about 68% of the heights are between $173 - 7 = 166$ and $173 + 7 = 180$ cm, about 95% are between $173 - 14 = 159$ and $173 + 14 = 187$ cm, and about 99.7% are between $173 - 21 = 152$ and $173 + 21 = 194$ cm.

Unbiasedness of s^2. If $x_1, x_2, \ldots, x_n$ are a random sample from a distribution, then the statistic s^2 computed with a divisor of $n - 1$ is an unbiased estimate of the true variance σ^2 of the distribution. This means that s^2 is correct on the average, in the sense that if s^2 were computed for all possible samples, and the mean of all these values were computed, that mean would be equal to the true value σ^2. This is true regardless of the parent distribution

(distribution sampled from), provided that its variance exists. The reason for dividing by $n - 1$ instead of n is related to the fact that $n - 1$ is the number of degrees of freedom of the deviations from the mean. In fact, it would make some sense to use a divisor of $n - 1$ rather than n also in defining the MAD, but in this case the appropriate divisor depends upon the particular parent distribution.

Example 1.2 Central tendency and variability for daily RORs

Here are four weeks (twenty trading days) of rates of return (RORs) (%) of a stock (Allstate Insurance Co. common stock in the four weeks starting with Monday, 7-Oct-2002.)

$$-1.8, +1.4, -3.4, +4.8, +3.3,$$
$$-0.1, +2.9, +1.1, +6.1, +1.4,$$
$$+2.1, -0.7, +0.8, -2.7, +0.6,$$
$$-0.6, -1.8, +1.2, -0.5, -1.1.$$

The mean of the RORs is

$$\frac{(-1.08) + 1.4 + (-3.4) + \cdots + (-1.1)}{20} = +0.65\%.$$

The order statistic of the RORs is
$$-3.4, -2.7, -1.8, -1.8, -1.1, -0.7, -0.6, -0.5, -0.1, +0.6,$$
$$+0.8, +1.1, +1.2, +1.4, +1.4, +2.1, +2.9, +3.3, +4.8, +6.1.$$
Choose convenient intervals ("bins") into which to place the numbers, say

$$(-4, -2), (-2, 0), (0, +2), (+2, +4), (+4, +6), (+6, +8).$$

The frequencies in these intervals are   2, 7, 6, 3, 1, 1. The cumulative frequencies are 2, 9, 15, 18, 19, 20. The cumulative relative frequencies are .10, .45, .75, .90, .95, 1.00. The upper quartile corresponds to a cumulative relative frequency of .75, so it is $+2\%$. The lower quartile corresponds to a cumulative relative frequency of .25, so it is in the second interval, $(-2, 0)$. It is approximated by interpolation as follows. The value -2 corresponds to a cumulative relative frequency of .10; the value 0, to a cumulative relative frequency of .45. Hence, $Q_1 \approx -2 + [(.25 - .10)/(.45 - .10)](2) = -2 + (15/35)(2) = -2 + .85 = -1.15\%$. Similarly, the median lies in the third interval, $(0, +2)$, and

$$Q_2 \approx 0 + [(.50 - .45)/(.75 - .45)](2) = (.05/.30)(2) \approx +0.08\%.$$

The minimum is -3.4%; the maximum, $+6.1$ %. The five-number summary, consisting of the minimum, lower quartile, median, upper quartile, and maximum, is

$$
\begin{aligned}
\text{minimujm} &= -3.4 \\
\text{lower quartile} &= -1.15 \\
\text{median} &= +0.08 \\
\text{upper quartile} &= +2.0 \\
\text{maximum} &= = +6.1
\end{aligned}
$$

The IQR is $+2.0 - (-1.15) = 3.15$ %. The mean is about $+0.65\%$. The deviations from the mean are

$$
-4.05, -3.35, \ldots, 5.45.
$$

The distances from the mean are

$$
4.05, 3.35, \ldots, 5.45.
$$

The standard deviation is the RMS of the distances, with a divisor of $n-1$ and comes out to about 2.42%. Similar steps are used to find the lower quartile Q_1 and the second quartile Q_2, which is the median.

In finding the quartiles, linear interpolation was used. Given (x_1, y_1) and (x_3, y_3), the problem is to approximate the value y_2 corresponding to a given value x_2 between x_1 and x_3. It proceeds as follows:

$$
\frac{y_2 - y_1}{y_3 - y_1} \approx \frac{x_2 - x_1}{x_3 - x_1}
$$

Then

$$
y_2 \approx y_1 + (y_3 - y_1)\frac{x_2 - x_1}{x_3 - x_1}.
$$

Linear interpolation is exact if the relationship between y and x is linear. For, then, there exist constants a, b such that for any x, the corresponding y is given by $y = a + bx$. In particular, this is true for $(x_1, y_1), (x_2, y_2), (x_3, y_3)$. That is,

$$
\frac{y_2 - y_1}{y_3 - y_1} = \frac{x_2 - x_1}{x_3 - x_1},
$$

so

$$
y_2 = y_1 + (y_3 - y_1)\frac{x_2 - x_1}{x_3 - x_1},
$$

exactly.

1.5   Higher Moments

The k-th *moment* is $\mathcal{E}[x^k]$. The k-th *central moment* is $\mathcal{E}[(x - \mu_x)^k]$. The *variance* is the second central moment.

Normalized versions of the third and fourth central moments are often considered as measures of *skewness kurtosis* (peakedness) of a distribution. For symmetric distributions, the positive and negative deviations cancel out one another, so odd moments are zero. A negative skew indicates that the left tail of the distribution is longer than the right side and many of the values (possibly including the median) lie to the right of the mean. A positive skew indicates that the right tail is longer than the left side and many of the values lie to the left of the mean. For a distribution of a r.v. X,   the skewness is $\mathcal{E}[[(X - \mu)/\sigma]^3] \;=\; \mu_3/\sigma^3$. The fourth standardized moment is μ_4/σ^4. For a Normal distribution, this is $3\sigma^4/\sigma^4 \;=\; 3$. So kurtosis is commonly defined as $\mu_4/\sigma^4 - 3$.

1.6   Summarizing Distributions*

*(Sections marked with * are more advanced or not in the mainstream of the development and may be considered as optional.)*

This section considers the topics of partitioning distributions and the so-called moment-preservation method. It discusses numerical summarization of distributions and datasets by discrete-distribution approximation and by partitioning.

1.6.1   Partitioning Distributions*

Cox (1957) gave recommendations for summarizing a distribution by a few intervals. For example, for the standard Normal for three groups the probabilities should be .27, .46, .27 to minimize the within-groups sum of squares. This is called the *twenty-seven percent rule*.   The two boundaries are approximately -0.61 and $+0.61$; that is, the intervals are $(-\infty, -0.61), (-0.61, +0.61), (+0.61, \infty)$. The corresponding group means are approximately $-1.22, 0$, and approximately $+1.22$. See Johari and Sclove (1976) for further discussion.

Partitioning a sample according to the minimum within-groups sum of squares leads to the *K-means algorithm* (MacQueen 1967) and *ISODATA* (Ball and Hall 1965, 1967). These algorithms are particularly interesting for multivariate data. The integer K denotes the number of clusters to be tried. One can try a range of values of K, one at a time.

ISODATA starts with initial guesses of K centers, assigns each observation to the nearest center, updates each center as the mean vector of the cases assigned to it, and so forth, continuing until convergence. K-means performs an update after each case is assigned. These algorithms converge to a minimum-distance partition of the set of observations.

1.6.2 Moment-Preservation Method*

Another way to summarize a distribution, be it a data distribution or a theoretical one, is to make a discrete approximation to it with K mass points $c_1, c_2, \ldots, c_K$ and their probabilities $p_1, p_2, \ldots, p_K$, by matching moments. The moments of the sample are $m'_r = \sum_{i=1}^{n} x_i^r / n$, $r = 1, 2, \ldots$. The first moment of a sample is $m'_1 = m = \bar{x}$. The method of summarizing based on matching moments is called the *moment-preservation method*. See Harris (2002). The method can be applied to moments m'_r of a sample or moments μ'_r of a probability distribution.

The unknowns whose values are to be found are K mass points and K probabilities; among these there are only $2K - 1$ free unknowns because the probabilities sum to 1. For example, with $K = 2$, the equations are $p_1 c_1 + p_2 c_2 = m'_1$, $p_1 c_1^2 + p_2 c_2^2 = m'_2$, $p_1 c_1^3 + p_2 c_2^3 = m'_3$, $p_1 + p_2 = 1$. The *central moments* of the sample are

$$m_r = \sum_{i=1}^{n} (x_i - \bar{x})^r / n, \ r = 1, 2, \ldots;$$

m_r is called the r-th central moment of the sample. (These are often modified to give unbiased estimates, as in the case $r = 2$ when a divisor $n - 1$ is used instead of n.) The equations can be re-expressed in terms of the central moments. For example, the second and third central moments in terms of the raw moments are $m_2 = m'_2 - m^2$ and $m_3 = m'_3 - 3m'_2 m + 2m^3$, respectively.

The moments of a two-valued variable Y taking values c_1, c_2 with probabilities p_1, p_2 can be found from those of a binary $(0,1)$ variable, that is, a variable which takes the value 0 with probability p_1 and the value 1 with probability p_2. Such a variable is called a *Bernoulli variable;* denote it by B. Its mathematical expectation is $\mathcal{E}[B] = 0 \cdot p_1 + 1 \cdot p_2 = p_2$; in fact, $\mathcal{E}[B^k] = 0^k \cdot p_1 + 1^k \cdot p_2 = 0 \cdot p_1 + 1 \cdot p_2 = p_2$. The variance is $\mathcal{V}[B] = (0 - p_2)^2 p_1 + (1 - p_2)^2 p_2 = p_1 p_2^2 + p_1^2 p_2 = p_1 p_2 (p_2 + p_1) = p_1 p_2 (1) = p_1 p_2$. Now, $Y = c_1$ when $B = 0$ and $Y = c_2$ when $B = 1$, so $Y = c_1 + (c_2 - c_1)B$. The mean of Y is $\mathcal{E}[Y] = \mathcal{E}[c_1 + (c_2 - c_1)B] = c_1 + (c_2 - c_1)\mathcal{E}[B] = c_1 + (c_2 - c_1)p_2 = p_1 c_1 + p_2 c_2$. The variance of Y is $\mathcal{V}[Y] = \mathcal{V}[c_1 + (c_2 - c_1)B] = \mathcal{V}[(c_2 - c_1)B] = (c_2 - c_1)^2 \mathcal{V}[B] = (c_1 - c_2)^2 p_1 p_2$. The third central moment of B is $m'_3 - 3m'_2 m'_1 + 2m_1^{'3} = p_2 - 3p_2^2 + 2p_2^3 = p_2(1 - 3p_2 + 2p_2^2) = p_2(1 - p_2)(1 - 2p_2) = p_2 p_1 (p_1 + p_2 - 2p_2) = p_1 p_2 (p_1 - p_2)$.

The third central moment of $Y = c_1 + (c_2 - c_1)B$ is $(c_2 - c_1)^3$ times that of B, or $(c_2 - c_1)^3 p_1 p_2 (p_1 - p_2)$.

Setting up the problem in terms of matching second and third central moments gives

$$p_1 c_1 + p_2 c_2 = m$$
$$p_1 p_2 (c_1 - c_2)^2 = m_2$$
$$p_1 p_2 (p_1 - p_2)(c_2 - c_1)^3 = m_3$$
$$p_1 + p_2 = 1.$$

The *skewness* is the normalized third moment, $m_3/m_2^{3/2}$. It is invariant under linear transformation, so the skewness of Y equals the skewness of B. Thus the skewness parameter is $p_1 p_2 (p_1 - p_2)/(p_1 p_2)^{3/2} = (p_1 - p_2)/(p_1 p_2)^{1/2}$. This theoretical skewness $\gamma = (p_1 - p_2)/(p_1 p_2)^{1/2}$ is matched with the sample skewness $g = m_3/m_2^{3/2}$. That is, the equation $(p_1 - p_2)/(p_1 p_2)^{1/2} = g = m_3/m_2^{3/2}$ replaces the equation in m_3. Note that $p_1 - p_2$ and $p_1 p_2$ can be expressed in terms of one another, so that one can be eliminated from this expression. To see this, use the identity $xy = (1/4)[(x+y)^2 - (x-y)^2]$ with $x = p_1, y = p_2$. This gives $p_1 p_2 = (1/4)[(p_1 + p_2)^2 - (p_1 - p_2)^2] = (1/4)[1 - ([p_1 - p_2]^2]$ and $4 p_1 p_2 = 1 - (p_1 - p_2)^2$, or $(p_1 - p_2)^2 = 1 - 4 p_1 p_2$. This gives $g^2 = (p_1 - p_2)^2/(p_1 p_2) = (1 - 4 p_1 p_2)/(p_1 p_2) = 1/(p_1 p_2) - 4$. Hence $1/(4 + g^2) = p_1 p_2 = p_1 - p_1^2$, giving the quadratic $p_1^2 - p_1 + 1/(4 + g^2) = 0$. This is of the form $x^2 - x + c = 0$ with $x = p_1, c = 1/(4 + g^2)$. The roots are $x = 1/2 \pm \sqrt{1/4 - c}$, that is, $p_1 = 1/2 \pm \sqrt{1/4 - 1/(4 + g^2)} = 1/2 \pm \sqrt{(1/4)[1 - 4/(4 + g^2)]} = 1/2 \pm 1/2 \sqrt{1 - 4/(4 + g^2)} = 1/2[1 \pm g \sqrt{\frac{1}{g^2 + 4}}]$. (See also Tabatabai and Mitchell 1984). Next, p_1 and p_2 can be replaced in the equations with their expressions in terms of g. The solution (Tabatabai and Mitchell 1984) is

$$c_1 = m - s\sqrt{p_2/p_1}, \quad c_2 = m + s\sqrt{p_1/p_2},$$

where s is the standard deviation, $s = \sqrt{m_2} = \sqrt{m_2' - m_1'^2}$, $p_2 = [1 + g\sqrt{\frac{1}{g^2 + 4}}]/2$, $p_1 = 1 - p_2$. Lin and and Tsai (1994) give a generalization to two variables and an example for three variables.

A result for $K = 3$. For $K = 3$ mass points, there are five free unknowns so the set of equations becomes more involved. However, it is interesting to consider a particular simple case: If the moment-preservation method is applied to a distribution, such as the standard Normal distribution, with mean $\mu = 0$, variance $\mu_2 = 1$, and third central moment $\mu_3 = 3$, the result is probabilities of $p_1 = 1/6, p_2 = 4/6, p_3 = 1/6$, on the points $c_1 = -\sqrt{3} \approx -1.732, c_2 = 0, c_3 = +\sqrt{3} \approx +1.732$. It is interesting to compare and contrast this with the results of partitioning the standard Normal distribution (Cox 1957, Johari and Sclove 1976), which gives probabilities of .27, .46, .27 and group means of $-1.22, 0, +1.22$.

1.7 Bivariate Data

Consider a pair of variables X and Y, such as height and weight.

1.7.1 Covariance and Correlation

The *covariance* of X and Y is $\mathcal{E}[(X - \mu_x)(Y - \mu_y)]$, that is, the covariance is the expected value of the cross-product of deviations from the mean. It is denoted also by $\mathcal{C}[X, Y]$.

The *correlation coefficient,* or, more simply, the correlation, of variables x and y denoted by ρ_{xy}, is their covariance, divided by the product of their standard deviations:

$$\text{Corr}(X, Y) = \frac{\mathcal{C}[X, Y]}{\text{SD}[x]\text{SD}([y]},$$

that is

$$\rho_{xy} = \frac{\sigma_{xy}}{\sigma_x \sigma_y},$$

where σ_x or $\text{SD}[x]$ is the standard deviation of x, and similarly for y.

The correlation is a *dimensionless* (unitless) quantity. The units in its numerator are cancelled by the units in its denominator. The range of the correlation coefficient is from $-1 \, to + 1$.

By reversing the above formula, the covariance is expressed as the product of the correlation and the standard deviations by reversing the above formula, that is,

$$\mathcal{C}[x, y] = \text{Corr}[x, y]\,\text{SD}[x]\,\text{SD}[y],$$

or

$$\sigma_{xy} = \rho_{xy}\sigma_x\sigma_y.$$

Consider a sample of pairs $(x_i, y_i), i = 1, 2, \ldots, n$. The *sample covariance* is denoted by s_{xy}. It is

$$s_{xy} = \frac{1}{n-1}\sum_{i=1}^{n}(x_i - \bar{x})(y_i - \bar{y}).$$

Consider the contribution of Case i to the covariance, the i-th term $(x_i - \bar{x})(y_i - \bar{y})$. Being a product of two factors, it is positive if both have the same sign and negative if they have opposite signs. So this is positive if both x_i and y_i are above average and if both are below average; it is negative if one is above average and the other is below average. The covariance is the average of such contributions (except the usual divisor is $n-1$ rather than n.)

1.7.1.1 Computational Formulas

The numerator of the sample covariance is the sum of products of deviations for x and y. Here are alternative equivalent expressions for the sum of products of deviations:

$$
\begin{aligned}
\sum (x_i - \bar{x})(y_i - \bar{y}) &= \sum (x_i - \bar{x})y_i \\
&= = \sum x_i(y_i - \bar{y}) \\
&= \sum x_i y_i - n\bar{x}\bar{y} \\
&= \sum x_i y_i - (1/n)\left(\sum x_i\right)\left(\sum y_i\right).
\end{aligned}
$$

1.7.1.2 Covariance, Regression Cooefficient, and Correlation Coefficient

The covariance is related to the regression coefficient and to the correlation coefficient; the covariance is the numerator of both. The coefficient of x in the regression of Y on x is s_{xy}/s_x^2. The sample correlation coefficient is

$$
\hat{\rho}_{xy} = \frac{s_{xy}}{s_x s_y}.
$$

(Here correlation is denoted by the symbol $\hat{\rho}$ rather than the frequently used symbol r because the often-used symbol r denotes continuous rate of return in this book.) Regression and correlation will be studied in some detail in later chapters.

1.7.2 Covariance of a Bivariate Distribution

The covariance of a bivariate distribution, that is, of random variables X and Y, is denoted by $C[x, y]$ or σ_{xy}. If X and Y are discrete with values v_j, $j = 1, 2, \ldots, J$ and w_k, $k = 1, 2, \ldots, K$, and joint probability mass function $p_{X,Y}(v_j, w_k) = \Pr\{X = v_j, Y = w_k\}$, then

$$
\sigma_{xy} = \sum_{j=1}^{J} \sum_{k=1}^{K} p_{X,Y}(v_j - \mu_x)(w_k - \mu_y).
$$

If X and Y are continuous with joint probability density function $f_{X,Y}(v, w)$, then $\sigma_{xy} = \int \int (v - \mu_x)(w - \mu_y) f_{X,Y}(v, w)\, dv\, dw$.

It can be shown that the covariance of the sum and the difference of two random variables is zero if and only if their variances are equal. (See the exercises.) Thus a test for equality of variances of x and y when they are not independent uses the covariance of the sum and the difference. The test is the "Pitman–Morgan" test for equality of variances when, as for paired data, X

and Y are not independent. (See Morgan (1939), Pitman (1939); and for more recent reviews, see Harris (1985) and Piepho (1997).)

Suppose that we were trying to develop a test for equality of variances in the situation in which we have (x, y) pairs, $\{(x_i, y_i), i = 1, 2, ..., n\}$. Without trying to think about anything fancy, we might consider the difference between the two sample variances, $s_x^2 - s_y^2$. Then we might actually try comparing the contributions to the variances of x and y, within cases, that is, for each i, namely,

$$(x_i - \bar{x})^2 - (y_i - \bar{y})^2, \ i = 1, 2, \ldots, n.$$

Then we might notice, from $a^2 - b^2 = (a+b)(a-b)$, that this contribution of Case i to the difference in variances is

$$
\begin{aligned}
(x_i - \bar{x})^2 - (y_i - \bar{y})^2 &= [(x_i - \bar{x}) + (y_i - \bar{y})][(x_i - \bar{x}) - (y_i - \bar{y})] \\
&= [(x_i + y_i) - (\bar{x} + \bar{y})][(x_i - y_i) - (\bar{x} - \bar{y})] \\
&= (s_i - \bar{s})(d_i - \bar{d}),
\end{aligned}
$$

where $s_i = x_i + y_i$ and $d_i = x_i - y_i$. Summing (and dividing by $n-1$), we get the covariance of the sum and difference. The covariance of the sum and the difference is zero if and only if the variances are equal. So, a test of equality of variances is equivalent to a test of nullity of the covariance of the sum and difference. This test can be carried out by computing their correlation r_{sd} and then $t = \sqrt{N - 2}\, r_{sd}/\sqrt{1 - r_{sd}^2}$.

1.8 Three Variables

1.8.1 Pairwise Correlations

Consider three variables X, Y, Z. There are three pairwise correlations, $\rho_{xy}, \rho_{xz}, \rho_{yz}$. The correlation matrix must be non-negative definite. Therefore, its determinant must be non-negative. So not all triplets of correlations are possible. For example, if ρ_{yx} and ρ_{zx} are high and positive, then the third correlation ρ_{xy} cannot be negative and of large size.

1.8.2 Partial Correlation

The *partial correlation* of y and z, adjusting for x, is denoted by $\rho_{yz\cdot x}$. It is defined as the ordinary correlation of the residuals of y and z from their respective regressions on x. More precisely, let $\hat{y} = \alpha_{y\cdot x} + \beta_{y\cdot x} x$, $\hat{z} = \alpha_{z\cdot x} + \beta_{z\cdot x} x$, where these are the regressions of y and z on x. Define the residual variables $\tilde{y} = y - \hat{y}$ and $\tilde{z} = z - \hat{z}$. Then $\rho_{yz\cdot x} = \rho_{\tilde{y}\tilde{z}}$.

From this definition it can be shown that

$$\rho_{yz \cdot x} = (\rho_{yz} - \rho_{yx}\rho_{xz})/\sqrt{1 - \rho_{yx}^2}\sqrt{1 - \rho_{xz}^2}.$$

This may be interpreted as follows. The correlation ρ_{yz} is the total correlation between y and z. The product $\rho_{yx}\rho_{xz}$ is the strength of the indirect relationship between y and z, via x. The partial correlation is the difference of these, normalized.

Conditions on triplets of correlations can be found from the condition that the size of the partial correlations must be less than or equal to 1.

Notions of sets of correlations, correlation matrices, and partial correlations extend to more than three variables. See, for example, Anderson (2003) or Johnson and Wichern (2008).

1.9   Two-Way Tables

This section concerns looking at data via two sets of categories, simultaneously. A *two-way table* is the resulting tabulation. The cell contents may be measurements or counts.

Example 1.3 Production by day of week and shift

Production may vary by day of week and shift. A two-way table may assist in studying this variation. Managerial intervention may be warranted.

TABLE 1.2
Production (Number of Widgets), by Day of Week and Shift Simultaneously

		Shift			
		1st	2nd	3rd	Mean
	M	100	150	200	150
Day	T	190	200	210	200
of	W	210	200	190	200
Week	R	160	240	200	200
	F	200	220	180	200
	MEAN	172	202	196	190

The *lines* of the table are its *rows* and *columns*. The rows are the horizontal lines. The columns are the vertical lines. The *body* of the table is the number inside, fifteen in the case above. There is a right-hand *margin* and a bottom margin. The margins contain means or totals. The *stubs* of a table are the labels across the top and down the left side.

Let y_{ds} be the value for day d and shift s. Let $y_{d\cdot}$ be the mean for day d, let $y_{\cdot s}$ be the mean for shift s and let $y_{\cdot\cdot}$ be the overall mean, the mean of all (fifteen) numbers in the body of the table. A way of describing the data is

$$
\begin{aligned}
y_{ds} &= \text{overall mean} + \text{day effect} + \text{shift effect} + \text{error} \\
&= y_{\cdot\cdot} + (y_{d\cdot} - y_{\cdot\cdot}) + (y_{\cdot s} - y_{\cdot\cdot}) + (y_{ds} - y_{d\cdot} - y_{\cdot s} + y_{\cdot\cdot}).
\end{aligned}
$$

The *fitted value* $\hat{y}_{ds}$ for day d and shift s is the overall mean + day effect + shift effect, or

$$
y_{\cdot\cdot} + (y_{d\cdot} - y_{\cdot\cdot}) + (y_{\cdot s} - y_{\cdot\cdot}) = y_{d\cdot} + y_{\cdot s} - y_{\cdot\cdot}.
$$

It is the value that might be expected for day d and shift s.

What value might have been expected for Friday's 3rd shift? A solution is

$$
\begin{aligned}
\text{Value expected} &= \text{Friday mean} + \text{3rd shift mean} - \text{Overall mean} \\
&= 200 + 196 - 190, \text{ or } 206 \text{ units.}
\end{aligned}
$$

How many fewer than this were produced on Friday's 3rd shift? Expected - actual $= 206 - 180 = 26$ units fewer than expected.

1.9.1   Two-Way Tables of Counts

Given categories $A_1, A_2, \ldots, A_r$ and $B_1, B_2, \ldots, B_c$, the data may be the frequency (count) n_{ij} of cases in categories A_i and B_j.

Example 1.4 Enrollment in Bus Stat I and Bus Stat II

The university offers an undergraduate business statistics sequence, Business Statistics I-II. What are the distributions across majors of students in the two courses? We tabulated the numbers of students by major in Bus Stat I in the Fall Semester and in Bus Stat II in the next semester and computed the row percents.

	ACTG	FIN	IDS	MGMT	MKTG	Total
Bus Stat I	121 (45%)	65 (24%)	30 (11%)	27 (10%)	27 (10%)	270 (100%)
Bus Stat II	59 (33%)	60 (33%)	30 (17%)	11 (6%)	20 (11%)	180 (100%)

Source: Hypothetical data, but consistent with real patterns.

It appears that about two-thirds of the students continue into the second course. Although 10% of those in the first course are MGMT majors, only 6% of those in the second course are.

1.9.2 Turnover Tables

Turnover tables are based on n persons observed at each of two times.

Example 1.5 Political polling

There are two candidates, A and B. Suppose we have 100 registered voters observed at two times. For example, we might asked 100 registered voters in September and again in October which candidate they prefer.

TABLE 1.3
100 Registered Voters, Interviewed in September and Again in October as to Preferred Candidate, A or B

		October A	B	Total
September	A	48 (80%)	12 (20%)	60 (100%)
	B	16 (40%)	24 (60%)	40 (100%)
Total		64 (64%)	36 (36%)	100 (100%)

Eighty percent of those for A remained for A; only sixty percent of those for B remained for B. In September, 60 of the 100 were for A; in October, 64. Suppose the same transition probabilities apply going forward from October to November; what numbers would then be expected in November?

Brand switching. The method of a turnover table can be applied to assessing loyalty (in the sense of repeat buying) to two competing brands A and B.

1.9.3 Seasonal Data

We think of the "seasons" as Summer, Fall, Winter, Spring, but, financially, seasonal data include quarterly data and monthly data.

1.9.3.1 Data Aggregation

In analyzing monthly retail sales data, there is a problem in that the months have different numbers of days. And, how might you deal with the effect of Easter, which sometimes falls in March and sometimes in April? A solution would be to aggregate into bi-monthly data, that is, six periods a year. Then both March and April will fall into the second of the six periods.

1.9.3.2 Stable Seasonal Pattern

A pattern across quarters may be consistent from year to year; then that pattern can be used for forecasting.

Example 1.6 Best Buy quarterly sales

The table shows Best Buy quarterly sales for several years. The units are megabucks (M$); for example, 1,606 means 1,606 millions (1.606 M$) or 1.606 billion.

TABLE 1.4
Best Buy Quarterly Sales (megabucks)

Year	Q1	Q2	Q3	Q4	Total
1998	1,606	1,793	2,106	2,852	8,357
1999	1,943	2,182	2,493	3,458	10,076
⋮	⋮	⋮	⋮	⋮	⋮
2006	6,959	7,603	8,473	12,899	35,934
Total	35,218	38,701	43,978	61,542	179,439
Percentage	19.6%	21.6%	24.5%	34.3%	100%

The percentages based on the Total row, $35{,}218/179{,}439 = 19.6\%$, $38{,}701/179{,}439 = 21.6\%$, $43{,}978/179{,}439 = 24.5\%$, and $61{,}542/179{,}439 = 34.3\%$, are used as the stable seasonal pattern, provided the pattern is reasonably consistent from year to year.

Stable seasonal pattern (Chen and Fomby 1999) will be discussed further in the chapter on time series analysis (Chapter 8).

1.10   Summary

Datasets can be characterized in terms of modes and ways.

Statistics *pl.* are descriptions, often numerical; statistics *sing.* is the body of methods used to analyze statistics *pl.*

Ratios and indices such as miles per gallon (kilometers per liter), consumer price index, and body mass index are data *derived* from more fundamental measurements.

A time series is generated by observing a single variable over time.

1.11   Chapter Exercises

1.11.1   Applied Exercises

1.13 Best Buy's sales of 35,934 for the year 2006 were up 16.5% from 2005. Use this percentage to forecast total sales for 2007.

1.14 (continuation) Next, forecast sales for each quarter of 2007 using the seasonal pattern of 19.6%, 21.6%, 24.5%, and 34.3% across quarters.

1.15 (continuation) The sales for the first three quarters of 2007 were 7,927, 8,750, and 9,928. Compare your predictions with these actual figures.

1.16 Suppose that a town's maximum daily temperatures(degrees Fahrenheit) for June were six days in the 90s, ten days in the 80s, two days in the 70s, and twelve days in the 60s. Why is the mean or median not a good summary of the the month's daily high temperatures?

1.17 (continuation) Make a histogram of the data in the preceding exercise.

1.18 What is the standard deviation of the sample $x_1 = -1, x_2 = 0, x_3 = +1$?

1.19 What is the standard deviation of the sample $x_1 = -2, x_2 = 0, x_3 = +2$?

1.20 Download Best Buy quarterly sales and bring the table up to date. Is a stable seasonal pattern continuing?

1.21 Two-way table of counts. A sample of 1,600 registered voters were asked in September and October which candidate they preferred, C or D. The results are shown in Table 1.5. Compute the row percents and interpret the shift, if any.

TABLE 1.5
Preferred Candidate, C or D, in September and October

| | | Oct. | | |
		C	D	Total
Sept.	C	580	420	900
	D	270	330	700
Total		850	750	1,600

1.11.2 Mathematical Exercises

1.22 Express $C[X+Y, X-Y]$, the covariance of the sum and the difference of two variables, in terms of their variances. *Hints.* (i) $C[A_1+A_2, B_1+B_2] = FOIL = first + outer + inner + last = C[A_1, B_1] + C[A_1, B_2] + C[A_2, B_1] + C[A_2, B_2]$. (ii) $C[X, -Y] = -C[X, Y]$ (iii) $C[X, X] = VX]$.

1.23 Show that the covariance of the sum and the difference of two variables is zero if and only if their variances are equal.

1.24 Conditional expectation of a standard Normal variable. Suppose the random variable z has the standard Normal distribution. Show that the conditional p.d.f. of z, given $a < z < b$, is $\phi(z)/[\Phi(b) - \Phi(a)]$, where $\phi(z)$ is the p.d.f. and $\Phi(z)$ is the c.d.f.

1.25 Conditional expectation of a Normal variable. Show that if z has the standard Normal distribution, then the conditional expectation of z, given $a < z < b$, is $[\phi(b) - \phi(a)]/[\Phi(b) - \Phi(a)]$. Recall that $\phi(z) = (1\sqrt{2\pi}) \exp(-z^2/2)$. *Hint:* Integrate using the substitution $u = z^2/2$, $du = z\,dz$.

1.26 Bernoulli variable. If B has a Bernoulli distribution with parameter p, and $Y = -1$ if $B = 0$ and $Y = +1$ if $B = 1$, what is Y in terms of B? What are the mean and variance of B? What are the mean and variance of Y?

1.27 Third central moment. Show that $\mu_3 = \mu'_3 - 3\mu'_2\mu + 2\mu^3$.

1.28 Correlations of three variables. If $\rho_{xy} = .8$ and $\rho_{yz} = .9$, find a lower bound on the third pairwise correlation ρ_{xz} using the fact that the determinant of the correlation matrix must be positive.

1.29 (continuation) Derive the same lower bound using the fact that the size of the partial correlation $\rho_{xz \cdot y}$ must be less than 1.

1.12 Bibliography

Anderson, T. W. (2003). *An Introduction to Multivariate Statistical Analysis. 3rd ed.* John Wiley & Sons, Inc., New York. (First edition, 1958.)

Ball, Geoffrey H., and Hall, David J. (1965). ISODATA, a novel method of data analysis and pattern classification. Technical Report AD 699616, Stanford Research Institute, Menlo Park, CA.

Ball, Geoffrey H., and Hall, David J. (1967). A clustering technique for summarizing multivariate data. *Behavioral Science*, **12**, 153–155.

Carroll, J. Douglas, and Arabie, Phipps (1980). Multidimensional scaling. In *Annual Review of Psychology*, M. R. Rosenzweig and L. R. Porter (Eds.), **31**, 607–649.

Chen, Rong, and Fomby, Thomas (1999). Forecasting with stable seasonal pattern models with an application of Hawaiian tourist data. *Journal of Business and Economic Statistics*, **17**, 497–504.

Clemen, Robert T. (1996). *Making Hard Decisions: An Introduction to Decision Analysis.* Duxbury Press, Belmont CA. (1st ed. 1991.)

Clemen, Robert T., and Reilly, Terence (2004). *Making Hard Decisions, 2nd ed.* Southwestern College Publishing, Belmont, CA.

Cox, D. R. (1957). Note on grouping. *Journal of the American Statistical Association*, **52**, 543–547.

Golub, Andrew L. (1997). *Decision Analysis: An Integrated Approach.* (Paperback) John Wiley & Sons, New York, 1997.

Harris, Bernard (2002). The moment-preservation method of cluster analysis. Pages 98–103 in *Exploratory Data Analysis in Empirical Research: Proceedings of the 25th Meeting of the Gesellschaft für Klassification, 2001*, Manfred Schwaiger and Otto Opitz (Eds.), Springer-Verlag, Heidelberg-Berlin.

Harris, P. (1985). Testing for variance homogeneity of correlated variables. *Biometrika*, **72**, 103–107.

Johari, Shyam, and Sclove, Stanley L. (1976). Partitioning a distribution. *Communications in Statistics*, **A5**, 133–147.

Johnson, Richard A., and Wichern, Dean W. (2008). *Applied Multivariate Statistical Analysis. 6th ed.* Pearson Prentice Hall, Upper Saddle River, NJ.

Lin, Ja-Chen, and Tsai, Wen-Hsiang. Feature-preserving clustering of 2-D data for two-class problems using analytical formulas: An automatic and fast approach. *IEEE Transactions on Pattern Analysis and Machine Intelligence*, **16**, 554–560.

MacQueen, James B. (1967). Some methods for classification and analysis of multivariate observations. *Proceedings of the 5th Berkeley Symposium on Mathematical Statistics and Probability*, **1**, 281–297. University of California Press, Berkeley and Los Angeles, CA.

Morgan, W. A. (1939). A test for the significance of the difference between two variances in a sample from a normal bivariate population. *Biometrika*, **31**, 13–19.

Piepho, H.-P. (1997). Tests for equality of dispersion in bivariate samples – Review and comparison. *Journal of Statistical Computation and Simulation*, **56**, 353–372.

Pitman, E. J. G. (1939). A note on normal correlation. *Biometrika*, **31**, 1–12.

Stevens, S. S. (1966). Mathematics, measurement and psychophysics. In *Handbook of Experimental Psychology* (S. S. Stevens, Ed.). John Wiley & Sons, New York, 1–49.

Tabatabai, Ali J., and Mitchell, O. Robert (1984). Edge location to subpixel values in digital imagery. *IEEE Transactions on Pattern Analysis and Machine Intelligence*, **6(2)**, 188–201.

2

Stock Price Series and Rates of Return

CONTENTS

2.1 Introduction

This chapter introduces price series and rates of return of individual assets. Later chapters on portfolio analysis are concerned with *sets* of assets, such as a number of stocks, held by an investor.

A price series is, for example, the daily closing prices of a stock. The stock's daily rates of return are derived data, calculated from the price series.

2.1.1   Price Series

Stock data are available from such data services as CRSP, the Center for Research on Security Prices; WRDS, the Wharton Research Data Service; and on the Web at such sites as *finance.yahoo.com.*

Daily quotes include the open, high, low, and closing price for the day. In Table 2.1, the symbols O, L, H, C denote open, low, high, and closing price, respectively. From these, interday and intraday *rates of return* (RORs) can be computed. The interday (between-day) ROR is

$$(C_t - C_{t-1})/C_{t-1}.$$

The intraday (within-day) ROR is

$$(C_t - O_t)/O_t.$$

The intraday ROR would be analyzed for *day trading,* where a stock is bought at the open and sold at the close of the trading day.

TABLE 2.1
Format of Table of Stock Price Data

	Open	Low	High	Close
1	O_1	L_1	H_1	C_1
2	O_2	L_2	H_2	C_2
⋮	⋮	⋮	⋮	⋮
n	O_n	L_n	H_n	C_n

Stock closing prices are recorded at the end of successive time periods. The time period could be days, weeks, months, quarters, or years. Here, for definiteness, we talk in terms of months.

Here, notation such as C_1, C_2, C_3, ..., C_n will represent a *price series,* that is, the closing price of a share of Stock A at the end of successive months.

There is usually also a column at the right called Adjusted Close, which adjusts the Close for dividends, splits, and distributions. For example, if there is a 2-for-1 split, the prices before the date of the split are divided by 2, so that the time series does not have a jump simply because of the split.

In the financial context, past data are often called *historical data.* This use of the word "historical" does not have the same meaning as in everyday usage, where a state's historical records might mean its records going back to the beginning of its statehood.

2.1.2 Rates of Return

If the price per share of a stock went from \$50 at the end of one month to \$52 at the end of the next, then its *rate of return* (ROR) for that month was $100\% \times (52 - 50)/50 = 4\%$. If the share price of another stock went from \$100 to \$102 in a month, its monthly rate of return was $(102 - 100)/100 \times 100\% = 2/100 \times 100\% = 2.0\%$.

Portfolio analysis is based on available past data, for example, monthly RORs. Denote ROR_t by R_t. Usually RORs will be computed from closing prices C_t, but here the more general notation, P_t is used. The ROR is

$$R_t = \frac{P_t - P_{t-1}}{P_{t-1}}.$$

We call $P_t - P_{t-1}$ the *return* from $t-1$ to t; then the *rate of return* is the return, divided by P_{t-1}. Note that $R_t = P_t/P_{t-1} - 1$, that is, $P_t/P_{t-1} = 1 + R_t$.

2.1.2.1 Continuous ROR and Ordinary ROR

Also, the ordinary R_t is approximated by the *continuous ROR*, r_t, defined by $r_t = \ln P_t - \ln P_{t-1}$. To see this, note that $r_t = \ln(P_t/P_{t-1}) = \ln(1 + R_t) \approx R_t$, because $\ln(1 + x) \approx x$ for x close to 0, that is, $1 + x$ close to 1. Here, the approximation is close when P_t/P_{t-1} is close to 1; that is, when P_t is close to P_{t-1}. Also, the continuous ROR r_t is less than the ordinary ROR R_t : $r_t < R_t$; r_t is a sharp lower bound for R_t. A financial reason why the continuous ROR is less than the discrete ROR is that if the compounding is continuous, a smaller rate will produce the same gain.

2.1.2.2 Advantages of Continuous ROR

An advantage of continuous ROR over ordinary ROR is that continuous ROR is additive. For example, the sum of monthly RORs of a stock is its annual ROR. For months $t = 0, 1, \ldots, 12$, given prices P_t and writing $\ln P_t$ as p_t and $r_t = p_t - p_{t-1}$, we have

$$
\begin{aligned}
\text{Annual continuous ROR} \quad &= \quad p_{12} - p_0 \\
&= \quad (p_{12} - p_{11}) + (p_{11} - p_{10}) + \cdots + (p_1 - p_0) \\
&= \quad r_{12} + r_{11} + \cdots + r_1 \\
&= \quad \sum_{t=1}^{12} r_t.
\end{aligned}
$$

On the other hand, for ordinary ROR, the annual ordinary ROR equals

$$
\begin{aligned}
\frac{P_{12} - P_0}{P_0} &= [\frac{P_{12} - P_{11}}{P_{11}} P_{11} + \frac{P_{11} - P_{10}}{P_{10}} P_{10} + \cdots + \frac{P_1 - P_0}{P_0} P_0]/P_0 \\
&= (R_{12}P_{11} + R_{11}P_{10} + \cdots + R_1 P_0)/P_0 \\
&= \sum_{t=1}^{12} w_t R_t,
\end{aligned}
$$

a *weighted* average of the monthly ordinary RORs, with weights $w_t = P_{t-1}/P_0$.

Continuous RORs have another advantage over discrete RORs when used with seasonal data; namely, as above, the continuous RORs add up to the total; ordinary RORs do not. Daily and weekly RORs illustrate this in the next example.

Example 2.1 Daily data: Stock prices and RORs by day-of-the-week

Here, the RORs of the S&P Midcap 400 ETF (ticker symbol MDY) day by day for a couple of weeks in the year 2007 will be used to illustrate ordinary RORs and continuous RORs. The closing price on Friday, 20-April was 159.88. Then, for the next two weeks (ten days) the prices were, for M, T, W, R, F of the first week,   159.91,   160.00, 160.78, 161.68, 160.76, and for M, T, W, R, F of the second week,   158.44,   158.86, 160.70,   160.97, 162.06. The ten ordinary RORs are, using $\text{ROR}_t, R_t = (P_t - P_{t-1})/P_{t-1} = P_t/P_{t-1} - 1$,

$$
\begin{aligned}
R_1 &= 159.91/159.88 - 1 = -0.019\%, \\
R_2 &= 160.00/159.91 - 1 = 0.056\%, \\
R_3 &= 160.78/160.00 - 1 = 0.487\%, \\
R_4 &= 161.68/160.78 - 1 = 0.560\%, \\
R_5 &= 160.76/161.68 - 1 = -0.569\%, \\
R_6 &= 158.44/160.76 - 1 = -1.443\%, \\
R_7 &= 158.66/158.44 - 1 = 0.265\%, \\
R_8 &= 160.70/158.66 - 1 = 1.158\%, \\
R_9 &= 160.97/160.70 - 1 = 0.168\%, \\
R_{10} &= 162.06/160.97 - 1 = 0.677\%.
\end{aligned}
$$

The sum of these is $-0.019 + 0.056 + 0.487 + 0.560 - 0.569 - 1.443 + 0.265 + 1.158 + 0.168 + 0.677 = 1.379\%$, whereas the ordinary ROR for the two weeks is $(P_{10}/P_0 - 1 = 162.06/159.88 - 1 = 1.363\%$. The sum of the ordinary RORs is not equal to the ordinary ROR for the whole period. On the other hand, the sum of the continuous RORs r_t will be equal to the continuous ROR for the whole period. To see this, note that the continous RORs $r_t = \ln P_t - \ln P_{t-1} = \ln(P_t/P_{t-1})$ are

$r_1 = \ln(159.91/159.88) = -0.019\%,$
$r_2 = \ln(160.00/159.91) = 0.056\%,$
$r_3 = \ln(160.78/160.00) = 0.486\%,$
$r_4 = \ln(161.68/160.78) = 0.558\%,$
$r_5 = \ln(160.76/161.68) = -0.571\%,$
$r_6 = \ln(158.44/160.76) = -1.454\%,$
$r_7 = \ln(158.66/158.44) = 0.265\%,$
$r_8 = \ln(160.70/158.66) = 1.152\%,$
$r_9 = \ln(160.97/160.70) = 0.168\%,$
$r_{10} = \ln(162.06/160.97) = 0.675\%.$

The sum of these is $-0.019 + 0.056 + 0.486 + 0.558 - 0.571 - 1.454 + 0.265 + 1.152 + 0.168 + 0.675 = 1.354\%$.

With continuous RORs, one can combine the daily data into a total for the first week, a total for the second week, and means for M, T, W, R, F. See Table 2.2.

TABLE 2.2
Daily Continuous RORs (pct) for Two Weeks

	M	T	W	R	F	Total
Week 1	0.019	0.056	0.486	0.558	−0.571	0.549
Week 2	−1.454	0.265	1.152	0.168	0.675	0.806
Sum	−1.435	0.321	1.638	0.726	0.104	1.355
Mean	−0.717	0.160	0.819	0.363	0.052	0.135

The overall mean continuous ROR across the ten days was 0.135%.

Day-of-week effects. The price on Friday, April 20 was 159.88. The continuous ROR for the first week was $\ln 160.76 - \ln 159.88 = 0.549\%$; note that this is indeed equal to the total for Week 1. The continuous ROR for the second week was $\ln 162.06 - \ln 160.76 = 0.805\%$, which is indeed equal to the total for the row for Week 2. Only two weeks of data does not give very precise results, but for the daily means, note that T, W, R are okay but M and F are not so good. M was negative in Week 2 but positive in Week1; F was negative in Week 1 but positive in Week 2.

Computation with ordinary RORs. Ordinary RORs work multiplicatively rather than additively. Using ordinary ROR, to combine the daily results into a weekly ROR, the RORs are changed to multiplicative factors, as follows for the Week 2 data.

$$(1 + R_6) \times (1 + R_7) \times (1 + R_8) \times (1 + R_9) \times (1 + R_{10})$$
$$= (1 - 0.01443) \times (1 + 0.00265) \times (1 + 0.01158) \times (1 + 0.00168) \times (1 + 0.00677)$$

$$= 0.98557 \times 1.00265 \times 1.01158 \times 1.00168 \times 1.00677$$
$$= 1.00808 = (1 + 0.00808),$$

giving the weekly ordinary ROR for Week 2 as about 0.808%. This compares to $P_{10}/P_5 - 1 = 162.06/160.76 - 1 = 0.008087 = 0.8087\%$, a little different due to round-off error in the individual daily ordinary RORs. The continuous ROR for Week 2 is about 0.806%.

Another type of ROR is the *intraday ROR,* that is, the within-day ROR, comparing the day's closing price to its opening price, as $(C_t - O_t)/O_t$, or in the continuous form $\ln C_t - \ln O_t$. Intraday ROR is of interest in day-trading, to be discussed later. The prefix *intra-* means within; *inter-* means between. So, by way of contrast, the ROR based on closing prices for two successive days is $(C_t - C_{t-1})/C_{t-1}$, or $\ln C_t - \ln C_{t-1}$, the *interday ROR.*

2.1.2.3 Modeling Price Series

The exponential function. Define a function, to be called $\exp(x)$, as

$$\exp(x) = \lim_{n \to \infty} (1 + x/n)^n.$$

Then it will turn out that

$$\exp(x) = 1 + x + x^2/2 + x^3/3! + \cdots + x^n/n! + \ldots.$$

To see this, note that, by taking $a = 1$ and $b = x/n$ in the binomial expansion

$$(a+b)^n = \sum_{k=0}^{n} C(n,k)\, a^k b^{n-k},$$

where $C(n,k) = n!/k!(n-k)! = n(n-1)(n-2)\cdots(n-k+1)/k!$,

$$
\begin{aligned}
(1+x/n)^n &= 1 + n(x/n) + [n(n-1)/2](x/n)^2/2 \\
&\quad + [n(n-1)(n-2)]\,(x/n)^3/3! + \cdots + (x/n)^n \\
&= 1 + x + [n(n-1)/n^2]\,x^2/2 \\
&\quad + [(n/n)(1-1/n)(1-2/n)x^3/3! + \cdots + (x/n)^n.
\end{aligned}
$$

The result then follows by taking limits as $n \to \infty$.

It can further be shown that $\exp(x) = e^x$, that is, the values of the function $\exp(x)$ can be found by taking a particular number e to the power x. This proof proceeds as follows (Kemeny 1957). Give $\exp(1)$ the name e. Then it can be shown that $\exp(n) = e^n$ for integers n. Then for rationals m/n, the ratio of two integers, it is shown that $\exp(m/n) = e^{m/n}$ and then finally it follows that for all real numbers x, $\exp(x) = e^x$.

From interest rates to geometric Brownian motion. Suppose that a principal amount P_0 is held for one time period at an interest rate R. Then the principal amount P_1 at the end of the period is $P_1 = P_0(1+R)$. If the interest is compounded m times during the period, then $P_1 = P_0(1+R/m)^m$. If the period is a year, the number m might be 2 for semi-annual compounding, 4 for quarterly compounding, or 12 for monthly compounding. If the compounding is "continuous," at a rate r, then the interest factor is $\lim_{m\to\infty}(1+r/m)^m = e^r$.

Now consider n periods, with different rates $r_1, r_2, \ldots, r_n$. Then

$$P_n = P_0(1 + r_1/m)^m(1 + r_2/m)^m \ldots (1 + r_n/m)^m.$$

Then

$$\lim_{m\to\infty} (1 + r_1/m)^m(1 + r_2/m)^m \ldots (1 + r_n/m)^m = e^{r_1}e^{r_2}\ldots e^{r_n}$$

$$= \exp[\sum_{t=1}^{n} r_t]$$

This gives

$$P_n = P_0 \exp[\sum_{t=1}^{n} r_t].$$

Now, take the rates r_t to be random variables. Then $\{P_n, n = 1, 2, \ldots\}$ is a stochastic (random) process. If the r_t are independently Normally distributed, the stochastic process is called *geometric Brownian motion* (see, e.g., Ross 2010 or Parzen 1962, 1999), although *exponential* Brownian motion would be just as good a name.

Modeling the price series. Note that, letting $p_t = \ln P_t$, we have $p_t = (p_t - p_{t-1}) + (p_{t-1} - p_{t-2}) + \cdots + (p_1 - p_0) + p_0 = r_t + r_{t-1} + r_1 + p_0$. Exponentiating, $P_t = \exp(p_t) = \exp(r_t + r_{t-1} + \cdots + r_1 + p_0) = \exp(p_0) \exp(r_t + r_{t-1} + \cdots + r_1) = P_0 \exp[\sum_{u=1}^{t} r_u]$. This is an exponential (or geometric) process. When $r_1, r_2, \ldots, r_t, \ldots$ are independent and identically distributed according to a Normal distribution, then, as mentioned above, the time series $\{P_t\}$ is called *geometric Brownian motion*.

Often in analyzing a time series, one applies the log transform to the data if they vary over several orders of magnitude. Further, one takes the differences if the series is not level. Often this would be done from a purely statistical viewpoint. (This will be further discussed in the chapter on time series analysis.) So, analysis of the difference of the logs, that is, the continuous ROR, is suggested both by the financial and the statistical viewpoints.

Various statistics for RORs include the mean, variance, Sharpe ratio, and Value-at-Risk (VaR). These, will be discussed in this chapter, but discussion of these measures for several assets simultaneously (portfolios) is deferred to later chapters.

First, mean, variance will be reviewed, in terms of distributions and of data. (See also Chapter 1.)

2.1.3   Review of Mean, Variance, and Standard Deviation

2.1.3.1   Mean

The mean $\mathcal{E}[X]$ of a random variable X is the probability-weighted average of its possible values. It is also called the expected value or *mathematical expectation*. (The notation $\mathcal{E}$ is used because the expectation is a function of a *distribution*, not just a single variable.) The mean of a sample $x_1, x_2, \ldots, x_n$ is $\bar{x} = \sum_{i=1}^{n} x_i/n$.

2.1.3.2   Variance

The variance $\mathcal{V}[X]$ of a random variable X is $\mathcal{V}[X] = \mathcal{E}[(X - \mu_x)^2]$; that is, the variance is the expected value of the squared deviation from the mean. The variance is often denoted by σ_x^2.

The variance of a sample $x_1, x_2, \ldots, x_n$ is

$$s^2 = \sum_{i=1}^{n} (x_i - \bar{x})^2/(n-1).$$

2.1.3.3   Standard Deviation

The square root of the variance is the standard deviation, SD[X], or σ_x. The sample standard deviation s is the square root of the sample variance.

2.2   Ratios of Mean and Standard Deviation

Next, measures based on the mean and standard deviation together are discussed; these include the *coefficient of variation* and the *Sharpe ratio*.

2.2.1   Coefficient of Variation

The *coefficient of variation,* or *relative standard deviation,* is the ratio of the standard deviation to the mean, σ/μ for the distribution or $s/\bar{x}$ in the sample. Note that this ratio is a dimensionless quantity. The units are the same in the numerator and denominator, so they cancel out. If Stock A has a mean ROR of 18% per year and a standard deviation of ROR of 9% per year, the coefficient of variation is $9/18 = 0.5$. If Stock B has a mean ROR of 6% per year and a standard deviation of 2% per year, the coefficient of variation is $2/6 \approx 0.33$.

The reciprocal of the coefficient of variation is the mean divided by the standard deviation. This can be interesting for RORs. If Stock A has a mean ROR of 18% per year and a standard deviation of ROR of 9% per year, the ratio of mean to standard deviation is 2. If Stock B has a mean ROR of 6% per year and a standard deviation of 2% per year, the ratio of mean to standard deviation is 3.

2.2.2 Sharpe Ratio

Given the risk-free rate R_f for such an asset as Treasury bills, the *excess ROR* of a stock is $R - R_f$, the stock's rate of return R minus the risk-free rate. The *Sharpe ratio* (Sharpe 1966) of a stock with mean ROR μ and standard deviation of ROR equal to σ is the ratio of excess ROR to the standard deviation, $(\mu - R_f)/\sigma$. The Sharpe ratios of portfolios will be considered in the chapters on portfolio analysis.

2.3 Value-at-Risk

The *Value-at-Risk* (VaR) at level .05 is the value below which there is only a probability of .05. This is the fifth percentile of the probability distribution of ROR. The probability is .95 of having a value greater than the VaR. Here we study this concept in terms of ROR, R_p. Then the .05 VaR is defined by $\Pr\{R_p \leq \text{VaR}\} = .05$, that is, $\Pr\{R_p > \text{VaR}\} = .95$. More generally, the level α VaR, say VaR_α, is defined by $\Pr\{R_p \leq \text{VaR}_\alpha\} = \alpha$. It is the 100α-th percentile of the distribution of R_p.

2.3.1 VaR for Normal Distributions

Under an assumption that R_p has a Normal distribution, the VaR, like any percentile, depends only upon the mean μ and standard deviation σ. Then

$$
\begin{aligned}
\Pr\{R_p > \text{VaR}\} &= \Pr\{(R_p - \mu)/\sigma \\
&> (\text{VaR} - \mu)/\sigma\} \\
&= \Pr\{z > (\text{VaR} - \mu)/\sigma\} = .95,
\end{aligned}
$$

where z has the standard Normal distribution. But $\Pr\{z > -1.645\} = .05$. Therefore, we set $(\text{VaR} - \mu)/\sigma$ equal to -1.645. This gives $\text{VaR} = \mu - 1.645\sigma$ as the 95% value of VaR. For example, if μ is 0.5% per month and σ is 1 % per month, this is $\text{VaR} = 0.5 - 1.645(1) = -1.145\%$. It is unlikely (5% chance) that the ROR will be less than -1.145% per month.

The set of stocks having .05-level VaR less than a given value $\text{VaR} = v$

is the set having $\mu - 1.645\sigma \geq v$. In the (σ, μ)-plane, these are the stocks represented by the points above the line $\mu = v + 1.645\sigma$.

2.3.2   Conditional VaR

It is interesting to compute the conditional VaR, which is the conditional expectation of ROR, given that it exceeds some constant c. Note that if Z has the standard Normal distribution, then $\mathcal{E}[z \,|\, z > z_0] = \phi(z_0)/[1 - \Phi(z_0)]$, where $\phi(z)$ is the probability density function and $\Phi(z)$ is the cumulative distribution function. Now, if a rate of return R is distributed according to a Normal distribution with mean μ and standard deviation σ, then R is distributed as $\mu + \sigma z$, and it follows that

$$\mathcal{E}[R \,|\, R > c] = \mu + \sigma\phi[(c - \mu)/\sigma]/[1 - \Phi([(c - \mu)/\sigma].$$

VaR for portfolios will be considered later when portfolio analysis is discussed.

2.4   Distributions for RORs

The bell-shaped Normal distribution is a possibility for fitting distributions of RORs. However, Normal distributions drop off relatively rapidly as the variable becomes large or small. Sometimes the distribution of RORs has more weight in the tails than that; it is said to have "heavy" (or "fat") tails. Here is a way that could happen, starting with a Normal distribution. Suppose that $R_1, R_2, \ldots, R_n$ are independent and identically distributed (i.i.d.) according to a Normal distribution with mean zero, but the standard deviation σ_t varies with t. Put a distribution on σ, with p.d.f $f_\sigma(s)$, $0 < s < \infty$. The resulting marginal distribution of R will be a t distribution.

2.4.1 t Distribution as a Scale-Mixture of Normals

To develop this, let T and U be random variables, T given $U = u$ being distributed according to according to a chi-square distribution with m d.f., and the conditional distribution of T given $U = u$ being Normal with mean zero and variance m/u; that is, u/m is the reciprocal variance. If u is large, the conditional variance of T is small. The p.d.f. of U is

$$f_U(u) = c \, u^{m/2-1} e^{-u/2}, \; u > 0,$$

where $c = \Gamma(m/2) / 2^{m/2}$. One can then write the conditional p.d.f. of T, given that $U = u$, derive the (unconditional) p.d.f. of T as

$$f_T(t) = \int_0^\infty f_{T,U}(t,u)\,du = \int_0^\infty f_{T|U}(t|u)\,f_U(u)\,du,$$

and verify that it is a Student's t distribution, the p.d.f. of which is

$$f_T(t) = \text{Const.}\,(1 + t^2/m)^{-(m+1)/2},$$

where Const. $= \Gamma(\frac{m+1}{2}) / \sqrt{m\pi}\,\Gamma(\frac{m}{2})$.

This way of deriving the t distribution is summarized by saying that t distributions are *scale-mixtures* of Normal distributions.

The members of the family of t distributions have heavier tails than Normal distributions and so can be useful for modeling RORs in this case.

2.4.2 Another Example of Averaging over a Population

Putting a population distribution over a parameter can be a very helpful way of modeling. Here is another example. It is not specifically financial, but it is actuarial. It is a model for *accident rates* in a population. Suppose that the yearly number of accidents of any given individual i in a population is distributed according to a Poisson distribution with parameter λ_i accidents per year. Then the probability that an individual with parameter value λ has exactly k accidents in a year, $k = 0, 1, 2, \ldots$, is $e^\lambda \lambda^k / k!$. Some individuals are more accident prone (have a higher accident rate) than others, so different individuals have different values of λ. A distribution can be put on λ to deal with this. (See the exercises.)

2.4.3 Section Exercises

2.1 Assume that, in fact, over the population, λ has an exponential distribution with mean 1/2, $f(\lambda) = (1/2)\exp(-\lambda/2)$, $0 < \lambda < \infty$. What is the probability that a randomly selected individual has exactly k accidents in a year, $k = 0, 1, 2, \ldots$?

2.2 (continuation) What is the name of this distribution, and what is the parameter and its value in this case?

2.3 (continuation) Now suppose that λ has a Gamma distribution with shape parameter m and scale parameter γ,

$$f(\lambda; m, \gamma) = \text{Const.}\,\lambda^{m-1}\exp(-\lambda/\gamma).$$

(The preceding case was $m = 1$, $\gamma = 2$.) Find the value of Const. (The integral $\int_0^\infty t^{m-1} e^{-t}\,dt = \Gamma(m)$, the gamma function with argument m.)

2.4 (continuation) What is the probability that a randomly selected individual has exactly k accidents in a year, $k = 0, 1, 2, \ldots$?

2.5 (continuation) What is the name of this distribution, and what is the parameter and its value in this case?

2.6 Carry out the indicated details of the derivation of the t distribution as a scale mixture of Normal distributions.

2.7 If U has a chi-square distribution with m d.f., what is the p.d.f. of its reciprocal, $1/U$?

2.5   Summary

The daily report of a stock on an exchange includes its open, high, low, and closing prices.

The interday (between-day) ROR is the closing price on a given day, minus the closing price on the day before, divided by the latter. This is $(C_t - C_{t-1})/C_{t-1}$. The continuous interday ROR is $\ln C_t - \ln C_{t-1}$.

The intraday (within-day) ROR is the closing price on a given day, minus the opening price on that day, divided by the latter. This is $(C_t - O_t)//O_t$. The continuous intraday ROR is $\ln C_t - \ln O_t$.

The risk-free rate is taken as the rate of a risk-free asset such as three-month Treasury bills.

The *excess* ROR of a stock is the stock's rate of return minus the risk-free rate.

The *Sharpe ratio* of a stock is its mean excess ROR divided by the standard deviation of its ROR.

A stock's value at risk (VaR) at level *alpha* is the $100(1 - \alpha)$-th percentile of the ROR distribution of that stock.

Distributions of RORs may be heavy-tailed.

2.6   Chapter Exercises

2.8 Suppose that RORs for nine days are

$$-1.1, -1.2, +1.2, +0.1, -0.1, +0.2, +1.1, +0.9, +0.8\%.$$

Find the order statistic, median, lower quartile, upper quartile, min, and max. Give the five-number summary.

2.9 (continuation) Find the mean, deviations from the mean, distances from the mean, MAD, and standard deviation.

2.10 Suppose that RORs for ten days are

$$-1.1, -1.2, +1.2, +0.1, -0.1, +0.1, +0.2, +1.1, +0.9, +0.8\%.$$

Find the order statistic, median, lower quartile, upper quartile, min, and max. Give the five-number summary.

2.11 (continuation) Find the mean, deviations from the mean, distances from the mean, MAD, and standard deviation.

2.12 If a stock's share price goes down $10 a share, from $100 to $90, what is the rate of return? If now it goes up by $10 a share, what is the rate of return?

2.13 If a stock's share price goes from $90 a share to $100 a share, what is the discrete rate of return, and what is the continuous rate of return?

2.14 If the DJIA falls from a level of 12,000 points by 3%, what is now its level? If it now increases by 3%, what is its level? Answer: 11,989. By what percent would it have had to increase to be back at 12,000?

2.15 (continuation) Compute the continuous ROR of the change from 11,640 to 12,000.

2.16 The DJIA went up 490 points one day to close at 12,045. What was its close the previous day? What was its percent increase? What was the continuous ROR in percent?

2.17 Show that $\ln x \approx x - 1$ if x is close to 1. *Hint:* The first terms of a Taylor series expansion of $f(x)$ about $x = a$ imply that $f(x) \approx f(a) + (x - a)f'(a)$ when x is close to a. Take $f(x) = \ln x$ and $a = 1$.

2.18 Sharpe ratio If $\mu = 0.5\%$ per month, $R_f = 0.1\%$ per month, and $\sigma = 0.5\%$ per month, what is the Sharpe ratio?

2.19 If $\mu = 0.5\%$ per month and σ equals 1% per month, what is the 10% VaR?

2.20 VaR If μ is one-percent per month and σ is 1% per month, what is the 5% VaR?

2.21 If μ is 1% per month and σ is 1% per month, what is the 10% VaR?

2.22 (continuation) What then is the conditional VaR?

2.23 Show that as the number m of degrees of freedom tends to infinity, the p.d.f. of the t distribution converges pointwise to that of the standard Normal distribution.

2.7 Bibliography

Campbell, John Y., Lo, Andrew W., and MacKinlay, A. Craig (1997). *The Econometrics of Financial Markets.* Princeton University Press, Princeton, NJ.

Kemeny, John G. (1957). The exponential function. *American Mathematical Monthly,* **64,** 158–160.

Parzen, Emanuel (1962). *Stochastic Processes.* Holden-Day, San Francisco, CA.

Parzen, Emanuel (1999). *Stochastic Processes.* Reprint of Parzen (1962), Wiley Classics in Applied Mathematics, John Wiley & Sons, Inc., New York.

Ross, Sheldon M. (2010). *Introduction to Probability Models. 10th ed.* Academic Press (Elsevier), Burlington, MA.

Sharpe, William F. (1965). Mutual fund performance. *Journal of Business,* **39,** 119–138.

Shiryaev, A. N. (1999). *Essentials of Stochastic Finance–Facts, Models, Theory.* World Scientific Publishing, River Edge, NJ.

Taylor, S. J. (2005). *Asset Price Dynamics, Volatility, and Prediction.* Princeton University Press, Princeton, NJ.

2.8 Further Reading

Excellent, somewhat more advanced references on returns and their statistical properties include Campbell, Lo, and MacKinlay (1999), Shiryaev (1999), and Taylor (2005).

3

Several Stocks and Their Rates of Return

CONTENTS

3.1 Introduction

This chapter introduces price series and rates of return of several assets, that is, sets of assets. The price series from a single stock is a single time series; prices from several stocks constitute a *multiple time series*. That is, at time t, the prices $P_{1t}, P_{2t}, \ldots, P_{mt}$ of m assets are observed.

Stock closing prices are recorded at the end of successive time periods. The time period could be days, weeks, months, quarters, or years.

The RORs are, for $i = 1, 2, \ldots, m$ assets and $t = 1, 2, \ldots, n$ time periods, the ordinary ROR,

$$R_{it} = (P_{it} - P_{i,t-1}) / P_{i,t-1} = P_{it}/P_{i,t-1} - 1,$$

and the continuous ROR,

$$r_{it} = \ln(P_{it}/P_{i,t-1}) = \ln P_{it} - \ln P_{i,t-1}.$$

3.2   Review of Covariance and Correlation

The *covariance* of X and Y is $\mathcal{E}[(X - \mu_x)(Y - \mu_y)]$, that is, the covariance is the expected value of the cross-product of deviations from the mean. It is denoted also by $\mathcal{C}[X, Y]$.

The *correlation coefficient*, or, more simply, the correlation, of variables X and Y, denoted by ρ_{xy}, is their covariance, divided by the product of their standard deviations,

$$\mathrm{Corr}[X, Y] = \frac{\mathcal{C}[X, Y]}{\mathrm{SD}[X]\,\mathrm{SD}[Y]},$$

that is,

$$\rho_{xy} = \frac{\sigma_{xy}}{\sigma_x \sigma_y},$$

where σ_x or $\mathrm{SD}[X]$ is the standard deviation of X, and similarly for Y.

The correlation is a *dimensionless* (unitless) quantity. The units in its numerator are cancelled by the units in its denominator. The range of the correlation coefficient is from -1 to $+1$.

By reversing the above formula, the covariance is expressed as the product of the correlation and the standard deviations; that is,

$$\mathcal{C}[X, Y] = \mathrm{Corr}[X, Y]\,\mathrm{SD}[X]\,\mathrm{SD}[Y],$$

or

$$\sigma_{xy} = \rho_{xy}\sigma_x\sigma_y.$$

The *sample covariance* is denoted by s_{xy}. It is

$$s_{xy} = \sum_{i=1}^{n}(x_i - \bar{x})(y_i - \bar{y})/(n - 1).$$

The covariance is related to the regression coefficient and to the correlation coefficient; the covariance is the numerator of both. The coefficient of x in the regression of y on x is s_{xy}/s_x^2. The sample correlation coefficient is

$$\hat{\rho}_{xy} = \frac{s_{xy}}{s_x s_y}.$$

(Here we denote correlation by the symbol $\hat{\rho}$ because the often-used symbol r here denotes continuous ROR.) Regression and correlation will be studied in some detail in subsequent chapters.

Note that correlation $\mathrm{Corr}[X, Y]$, can apply to different r.v.s such as height and weight or also to r.v.s that are values of a single variable such as ROR at two different time points, such as $\mathrm{Corr}[R_t, R_{t-1}]$. This type of correlation is called *autocorrelation* and will be discussed in Chapter 8 on time series analysis.

Next, descriptive statistics for the cases of two, three, and m stocks are treated.

3.3 Two Stocks

3.3.1 RORs of Two Stocks

Consider two stocks, A and B. Represent their RORs by random variables X and Y, with means μ_x, μ_y and standard deviations σ_x, σ_y. The correlation is ρ_{xy}; the covariance $\sigma_{xy} = \rho_{xy} \sigma_x \sigma_y$. Table 3.1 gives RORs for two stocks, for n time periods.

TABLE 3.1
Format of Table of RORs for Two Stocks

Month	Stock A	Stock B
1	0.5 %	1.6 %
2	−0.3%	1.4 %
⋮	⋮	⋮
n	0.2 %	0.9 %

From the statistics in Table 3.2, it appears that stock A has a moderate ROR, and B has a higher ROR but is riskier. There is a negative correlation that is medium-large in size.

The probability that the ROR will be negative, the loss probability, is computed next for Stock A and Stock B.

If the variable X is taken as having a Normal distribution, what is the probability of loss with Stock A; that is, what is $\Pr\{X < 0\}$? The standardized value of $x = 0$ is $z = (0 - 0.4)/0.8 = -0.5$, so letting Z be a r.v. with the standard Normal distribution, $\Pr\{X < 0\} = \Pr\{Z < -0.5\} \approx .309$.

What is the probability of loss with Stock B; that is, assuming Normality, what is $\Pr\{(Y < 0\} =$? Here, $z = (0 - 1.5)/4.5 = -0.333$, and the probability to the left of that value under the standard Normal curve is about .370.

TABLE 3.2
Statistics of RORs of Two Stocks

Stock	A	B
ROR	x	y
Mean ROR	0.4 %	1.5 %
Std.Dev. of ROR	0.8 %	4.5 %
Correlation of RORs	−.6	

Example 3.1 The "January effect"

This effect is the tendency of security prices to tend to increase in January. However, here we view a *January effect* that is January RORs being more highly correlated with the annual ROR than are those of the other months. Here we are not viewing two different stocks but rather a stock index for January and for the whole year. The S&P500 monthly RORs for the sixteen years 1951 through 1966 gave the following correlations of each month with the annual ROR: J: .75, F: .11, M: .38, A: .48, M: .42, J: .35, J: .41, A: .41, S: .71, O, −.16, N: .15, D: .54. The ROR for January is highest (although September, at +.71, is also high). The January effect is something that might be taken into account early in the year in financial planning for the year ahead.

If the twelve months' RORs were uncorrelated and had the same variance and equal correlations with the annual ROR, then the squared value of that correlation would be $1/12$, corresponding to a correlation of $\sqrt{1/12} \approx .29$. To see this, let r_m be the ROR of month m and $r_+ = \sum_{t=1}^{m} r_t$ and write the covariance $\mathcal{C}[r_m, r_+] = \mathcal{C}[r_m, \sum_{t=1}^{12} r_t] = \sum_{t=1}^{12} \mathcal{C}[r_m, r_t] = \mathcal{C}[r_m, r_m] = \mathcal{V}[r_m]$. Then the correlation is $\mathrm{Corr}[r_m, r_+] = \mathcal{C}[r_m, r_+]/\mathrm{SD}[r_m]\mathrm{SD}[r_+] = \mathcal{V}[r_m]/\mathrm{SD}[r_m]\sqrt{12\mathcal{V}[r_m]} = 1/\sqrt{12} \approx .29$.

3.3.2 Section Exercises

3.1 Download recent monthly S&P500 data and compute the correlations of monthly RORs with the annual.

3.2 Download recent monthly DJIA data and compute the correlations of monthly RORs with the annual.

3.4 Three Stocks

3.4.1 RORs of Three Stocks

Next, consider three stocks, A, B, and C. Denote the RORs by X, Y, Z. The covariances are $\sigma_{xy} = \rho_{xy}\sigma_x\sigma_y$, $\sigma_{xz} = \rho_{xz}\sigma_x\sigma_z$, $\sigma_{yz} = \rho_{yz}\sigma_y\sigma_z$. The means, standard deviations, and correlations are estimated by their sample analogs, the sample means, sample standard deviations, and sample correlations.

In this example, the sample correlation of x and y is low and positive, that of x and z is low and positive, and that of y and z is negative and large in size.

TABLE 3.3
Format of Table of RORs of Three Stocks

Stock	A	B	C
ROR	x	y	z
Month 1	0.5%	1.6 %	−0.1%
Month 2	−0.3%	1.4 %	0.2%
$\vdots$	$\vdots$	$\vdots$	$\vdots$
Month n	0.2%	0.9 %	1.2%
Mean	0.4%	1.1%	0.9%
Std.Dev.	0.5%	3.1%	4.2%
Correlation Stock A Stock B		$\hat{\rho}_{xy} = +.2$	$\hat{\rho}_{xz} = +.3$ $\hat{\rho}_{yz} = -.7$

3.4.2 Section Exercises

3.3 Some triplets of correlations are not possible. In particular, the determinant of the correlation matrix must be positive. Verify that this is the case for the three correlations in Table 3.3.

3.4 Check to see whether the determinant of the correlation matrix formed from $\rho_{xy} = +.2$, $\rho_{xz} = +.3$, and $\rho_{yz} = -.9$ is positive.

3.5 Given $\rho_{xy} = +.2$ and $\rho_{xz} = +.3$, what range of values is possible for ρ_{yz}?

3.6 Check to see whether the determinant of the correlation matrix formed from $\rho_{xy} = +.2$, $\rho_{xz} = +.3$, and $\rho_{yz} = +.4$ is positive.

3.7 Given $\hat{\rho}_{xy} = +.8$ and $\rho_{xz} = +.8$, what range of values is possible for ρ_{yz}?

3.8 Partial correlation Another way to look at triplets of pairwise correlations is through the partial correlation. The partial correlation $\rho_{yz \cdot x}$ of y and z, adjusting for x, is defined as the ordinary correlation of the residuals of y and z from their respective regressions on x. From this definition it follows that $\rho_{yz \cdot x} = (\rho_{yz} - \rho_{yx}\rho_{xz})/\sqrt{1 - \rho_{yx}^2}\sqrt{1 - \rho_{xz}^2}$. Because the partial correlation is an ordinary correlation (of residual variables), its size must be less than or equal to 1. Use this fact as another way to do Exercise 3.7.

3.5 m Stocks

3.5.1   RORs for m Stocks

Next, consider m stocks, indexed by $i = 1, 2, \ldots, m$, with RORs $R_1, R_2, \ldots, R_m$, or, including the time t in the notation, $R_{1t}, R_{2t}, \ldots, R_{mt}$ at time t, $t = 1, 2, \ldots, n$.

3.5.2   Parameters and Statistics for m Stocks

The means are $\mu_i, i = 1, 2, \ldots, m$, the standard deviations are $\sigma_i, i = 1, 2, \ldots, m$, and the covariances are $\sigma_{ij} = \rho_{ij}\sigma_i\sigma_j$ for stocks $i, j = 1, 2, \ldots, m$. These parameters are estimated by their sample analogs, the sample means, sample standard deviations, and sample covariances $s_{ij}, i, j = 1, 2, \ldots, m$. All of this can be written and defined for continuous RORs r_{it} as well as ordinary RORs R_{it}.

3.6   Summary

Covariance is the expected value of the deviations of two r.v.s from their means.

Correlation is normalized covariance; that is, the correlation coefficient is

TABLE 3.4

Format of Table of RORs R_{it}, $i = 1, 2, \ldots, m$, $t = 1, 2, \ldots, n$

Stock	1	2	$\cdots$	m
ROR	R_1	R_2	$\cdots$	R_m
Month				
1	R_{11}	R_{21}	$\cdots$	R_{m1}
2	R_{12}	R_{22}	$\cdots$	R_{m2}
$\vdots$	$\vdots$	$\vdots$	$\cdots$	$\vdots$
n	R_{1n}	R_{2n}	$\cdots$	R_{mn}
Mean	$\bar{R}_1$	$\bar{R}_2$	$\cdots$	$\bar{R}_m$
Std.Dev.	s_1	s_2	$\cdots$	s_m

the covariance divided by the product of the standard deviations. The correlation coefficient is a dimensionless quantity; the units of the numerator equal the units of the denominator, so the units cancel out.

3.7 Chapter Exercises

3.9 Suppose that RORs of Stock A for five days are

$$-1.1, -1.2, +1.2, +0.1, -0.1\%$$

and RORs of Stock B for the same five days are

$$-1.2, -1.1, +1.1, +0.2, -0.2\%.$$

Find the covariance and correlation of the RORs of Stocks A and B.

3.10 Suppose that RORs of Stock A for nine days are

$$-1.1, -1.2, +1.2, +0.1, -0.1, +0.2, +1.1, +0.9, +0.8\%$$

and RORs of Stock B for the same nine days are

$$-1.2, -1.1, +1.1, +0.2, -0.2, +0.9, +1.1, +0.7, +0.9\%.$$

Find the covariance and correlation of the RORs of Stocks A and B.

3.11 If $r_{xy} = +.9$ and $r_{yz} = +.8$, find a non-trivial lower bound for r_{xz}.

3.12 If $r_{xy} = +.7$ and $r_{yz} = +.9$, find a non-trivial lower bound for r_{xz}.

.

3.8 Bibliography

Afifi, Abdelmonen, May, Susanne, and Clark, Virginia A. (2012). *Practical Multivariate Analysis. 5th ed.* CRC Press, Taylor & Francis Group, Boca Raton, FL.

Johnson, Richard A., and Wichern, Dean W. (2008). *Applied Multivariate Statistical Analysis. 6th ed.* Pearson Prentice Hall, Upper Saddle River, NJ.

3.9 Further Reading

The reader may wish to further review notions of covariance and correlation from basic books on statistics or business statistics, or books on applied multivariate analysis such as Afifi, May, and Clark (2012) and Johnson and Wichern (2008).

Part II

REGRESSION

4

Simple Linear Regression; CAPM and Beta

CONTENTS

4.1 Introduction

This chapter begins with an example of simple linear regression. Then the Capital Assets Pricing Model (CAPM), which may be viewed as a particular application of simple linear regression to finance, is considered. The CAPM relates the rate of return (ROR) of any given asset to the ROR of the market as a whole, indicated by a market index such as the S&P500 (Standard & Poor's 500 Stock Composite Index). It is shown how the CAPM characterizes each stock with a parameter *beta*. The CAPM is regression through the origin, with only a slope parameter. Next, the usual simple linear regression model with both slope and intercept is treated. The modification of the CAPM with intercept as well as slope is discussed.

4.2 Simple Linear Regression

The distribution of a variable Y may be studied in terms of its dependence on values x of another variable X. The word "simple" in the phrase *simple linear regression* refers to the fact that Y is being described in terms of a single explanatory variable X. In "multiple" regression, a topic of the chapter after this one, a response variable Y is described in terms of several explanatory variables.

When it is reasonable that $X = 0$ implies $Y = 0$, a model stating simply that Y is proportional to X can be appropriate. Then only a single parameter, the proportionality parameter, which is a slope, is needed. The conditional mean of Y, given that $X = x$, is $\mathcal{E}[Y \mid x] = \beta x$. The function $\eta(x) = \mathcal{E}[Y \mid x]$ is the *regression function*. In this case, it is simply $\eta(x) = \beta x$. The *statistical model* here is that the response variable Y is equal to its conditional expectation given x, plus a random deviation ε from that: $Y = \beta x + \varepsilon$.

4.2.1 Data

The data for simple linear regression take the form of (x, y) pairs for n cases or instances of observation, $(x_i, y_i), i = 1, 2, \ldots, n$.

The *observational model* is the statistical model, stated in terms of cases. This gives the *observational equations,* which here are

$$Y_i = \beta x_i + \varepsilon_i, \; i = 1, 2, \ldots, n.$$

The conditional mean of Y_i given x_i is βx_i. The variables ε_i are differences from the conditional mean and hence have mean 0.

4.2.2 An Introductory Example

An introductory example will be discussed in some detail. Suppose that for $n = 14$ trips of a car, the miles traveled (y) and gallons used (x) were recorded. The data are shown in Table 4.1. Figure 4.1 is a *scatterplot* of the data. The points are labeled with the numbers of the 14 runs.

TABLE 4.1
Gasoline Mileage Data

Run	Miles	Gallons	Run	Miles	Gallons
1	62	4.6	8	108	8.0
2	49	5.7	9	96	7.5
3	73	4.3	10	61	4.7
4	63	6.1	11	165	10.8
5	108	9.3	12	148	8.9
6	135	9.9	13	197	13.1
7	60	4.9	14	185	12.7

4.3 Estimation

Two estimates of the miles per gallon come to mind almost at once. One is simply the total miles divided by the total gallons, which is

$$\frac{\sum y_i}{\sum x_i}. \tag{4.1}$$

Note that

$$\frac{\sum y_i}{\sum x_i} = \frac{\sum y_i/n}{\sum x_i/n} = \frac{\bar{y}}{\bar{x}},$$

the *ratio of the means,* where

$$\bar{x} = \sum x_i/n \text{ and } \bar{y} = \sum y_i/n.$$

(The notation $\sum x_i$ is used as shorthand for $\sum_{i=1}^{n} x_i$ when the meaning is clear. Sometimes we abbreviate even further as $\sum x$.)

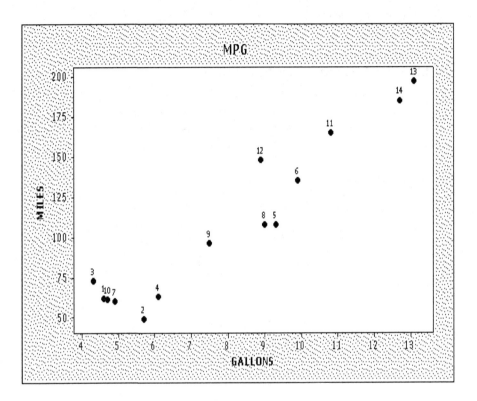

FIGURE 4.1
Miles versus gallons

Now, each ratio y_i/x_i is an estimate of the miles per gallon, so it also makes sense to consider an estimate that is the *mean of the ratios*,

$$\frac{1}{n} \sum \frac{y_i}{x_i}. \tag{4.2}$$

Note that the "ratio of means" estimate is equal to 13.67 mpg (miles per gallon). The mpg for the runs are shown. The mean of these fourteen ratios is

$$\text{Mean ratio} \; = \; (13.5 + 8.6 + \ldots + 14.6)/14 = 13.4 \text{ mpg}.$$

(To convert to a metric scale, note that one gallon is four quarts, and one liter is about 1.06 quarts (more precisely, 1.05668821). Also, one mile is 1.61 km. It follows that 1 km/liter = 2.34 mpg, or 1 mpg = 0.427 km/liter.)

TABLE 4.2
Miles per Gallon for Each of the Fourteen Runs

run	mpg	run	mpg
1	13.5	8	13.5
2	8.6	9	12.8
3	17.0	10	13.0
4	10.3	11	15.3
5	11.6	12	16.1
6	13.6	13	15.0
7	12.2	14	14.6

Later, the relative merits of the mean ratio and the ratio of the means will be discussed. There are still other approaches that should be explored.

One way to develop a model for the situation is to assume that the mean value function is of the form $\mathcal{E}[Y \mid x] = \beta x$, where the *regression coefficient* β, the miles per gallon, is an unknown parameter, to be estimated. The notation $\mathcal{E}[Y \mid x]$ denotes the *conditional* expectation of Y, given that $X = x$. If Y were weight and X were height, this would be the mean of the distribution of weights of men of given height x, for example, the mean weight of men 70 inches tall. The model implies that the mean weight of men 71 inches tall would be β pounds more than that of men 70 inches tall. (However, in this situation, an intercept as well as a slope would usually be used; this is considered later in the chapter.)

The model based on this conditional expectation states that the mean value of Y for any given value of x is *proportional* to that value: miles are proportional to gallons, β being the constant of proportionality. The graph of the function $y = \beta x$ is a line through the origin with slope equal to β. Hence estimating β in this model is called "fitting a line through the origin." For n instances of observation, the model is

$$Y_i = \beta x_i + \varepsilon_i, \ i = 1, 2, \ldots, n.$$

These n equations are called the *observational equations*. The variables

$$\varepsilon_i = Y_i - \mathcal{E}[Y_i \mid x_i] = Y_i - \beta x_i$$

are called the *errors*. This does not mean "error" in the sense of a mistake. The "error" is merely the departure from the (conditional) expected value. The errors are unobservable. However, if the value of β were somehow known, the values of the errors would be known, too, because βx_i could be subtracted from Y_i to give ε_i. (The error ε is a random variable, but it is denoted by a Greek letter because it is unobservable. Greek letters are used for unobservable quantitites, such as errors and parameters, and Latin letters are used for observable quantities.)

In the simplest case, it is asssumed that the errors are uncorrelated, have mean zero, and have common variance, denoted by σ^2. That ε_i and ε_j are uncorrelated is equivalent to $\mathcal{C}[\varepsilon_i, \varepsilon_j] = 0$ for $i \neq j$. Because the means of the errors are zero, this is $\mathcal{E}[\varepsilon_i \varepsilon_j] = 0$ for $i \neq j$. The assumption that the errors have common variance is written $\text{Var}(\varepsilon_i) = \sigma^2$, a constant not dependent upon i. These two assumptions may be combined conveniently as $\mathcal{C}[\varepsilon_i, \varepsilon_j] = \sigma^2 \delta_{ij}$, where δ_{ij} is the *Kronecker delta*, equal to 1 if $i = j$ and equal to 0 otherwise.

It turns out that when the errors do in fact have the same variance, a third estimator, different from the mean ratio or the ratio of means, will have a smaller variance than either of them. This third estimator is considered next. Later, when the mean ratio and ratio of means are considered further, it will be seen that the ratio of averages has the smallest variance when $\mathcal{V}[\varepsilon_i]$ is proportional to x, that is, is equal to $\sigma^2 x_i$, and the average of the ratios has the smallest variance when $\mathcal{V}[\varepsilon_i]$ is proportional to the square of x, *i.e.*, is equal to $\sigma^2 x_i^2$.

4.3.1 Method of Least Squares

4.3.1.1 Least Squares Criterion

Let b be any trial value for β, that is, any guess as to the value of the unknown parameter β. For $i = 1, 2, \ldots, n$, the value $\hat{y}_i(b) = bx_i$ is the predicted value of y_i corresponding to b. To assess the goodness of the choice of b for β, each observed value y_i is compared with its predicted value $\hat{y}_i$, using the error $y_i - \hat{y}_i$. To combine the results for all n cases, the errors are squared and summed. The result is

$$S(b) = \sum [y_i - \hat{y}_i(b)]^2 = \sum (y_i - bx_i)^2.$$

This is the *least squares criterion* for this problem. A reason that it is taken as the criterion is that it is the (square of) the distance between the points

$$P_1 : (y_1 \ y_2 \ \ldots \ y_n) \text{ and } P_2 : (\hat{y}_1 \ \hat{y}_2 \ \ldots \ \hat{y}_n)$$

in n-space.

4.3.1.2 Least Squares Estimator

The *principle of least squares* states that the choice of estimate of the parameter β is to be that value b^* of b that minimizes the least squares criterion. The quantity $S(b)$ is a score associated with the line $y = bx$; it is the sum of squared vertical deviations of the data points (x_i, y_i) from this line. A small score is thus a good score. The value of b that gives the line with the smallest score is the *least squares estimate* of the parameter β. This best value is denoted by b^*. It is given by the sum of products over the sum of squares of the values of x, that is, by the expression

$$b^* = \frac{\sum x_i y_i}{\sum x_i^2}. \tag{4.3}$$

To obtain this expression, note that

$$
\begin{aligned}
S(b) &= \sum [y_i - \hat{y}_i(b)]^2 \\
&= \sum (y_i - bx_i)^2 \\
&= \sum (y_i - 2bx_i y_i + b^2 x_i^2) \\
&= \sum y_i^2 - 2b \sum x_i y_i + b^2 \sum x_i^2 \\
&= \left(\sum x_i^2\right) b^2 + \left(-2 \sum x_i y_i\right) b + \sum y_i^2 \\
&= \quad A\, x^2 \quad + \quad B\, x \quad + \quad C,
\end{aligned}
$$

with $x = b$, $A = \sum x_i^2$, $B = -2 \sum x_i y_i$, and $C = \sum y_i^2$. The stationary point of such a quadratic is $x = -B/2A$. If $A > 0$, as here, the stationary point yields a minimum. Here the stationary point is $b = -(-2 \sum x_i y_i)/2 \sum x_i^2 = \sum x_i y_i / \sum x_i^2$. That is, the least squares estimator is $b^* = \sum x_i y_i / \sum x_i^2$. This result can be written succinctly as

$$b^* = \arg\min_b S(b),$$

where $S(b)$ is the least squares criterion $\sum_{i=1}^n (y_i - bx_i)^2$. This says that b^* is the value of the argument b of the function $S(b)$ that minimizes that function.

Summary statistics computed from the data in Table 4.1 are given in Table 4.3. For these data, the least squares estimate is

TABLE 4.3
Summary Statistics for Gasoline Mileage Data

		$n = 14$		
$\sum x = 110.5$ gallons				$\sum y = 1{,}512$ miles
$\sum x^2 = 988.95$				$\sum y^2 = 196{,}184$
		$\sum xy = 13{,}797.7$		

$$
\begin{aligned}
b^* &= \sum x_i y_i / \sum x_i^2 \\
&= 13{,}797.7 \text{ miles} \times \text{gallons} / 988.95 \text{ gallons}^2 \\
&= 13.952 \text{ miles per gallon (mpg)}.
\end{aligned}
$$

4.3.2 Maximum Likelihood Estimator under the Assumption of Normality*

Now assume that the errors ε_i are independent and jointly Normally distributed, each with mean zero and variance σ^2. The (conditional) probability density function of Y_i given x_i is

$$(2\pi\sigma^2)^{-1/2} \exp[-(y_i - \beta x_i)^2/2\sigma^2].$$

By independence, the joint probability density function is the product, which simplifies to

$$(2\pi\sigma^2)^{-n/2} \exp[-\sum(y_i - \beta x_i)^2/2\sigma^2].$$

The joint probability density function involves the parameters β and σ^2. It is considered a function of the observations, with the parameters fixed. On the other hand, the *likelihood* L is the joint probability density, considered a function of the parameters, with the observations fixed. The *method of maximum likelihood* consists of choosing as the parameter estimates those values that maximize the likelihood, that is, those parameter values that are most consistent with the observed values of the data, in the sense of maximizing their probability.

To emphasize that in the likelihood the parameters are variables, we write β as the variable b and σ^2 as the variable v:

$$L = (2\pi v)^{-n/2} \exp[-\sum(y_i - bx_i)^2/2v].$$

The *principle of maximum likelihood* says to choose as the estimates of β and σ^2 the values of B and v that make the likelihood largest.

Note that

$$L = (2\pi v)^{-n/2} \exp[-S(b)/2v],$$

so that, for any value v, L is maximized by minimizing $S(b)$. But the value of b that minimizes $S(b)$ is just the least squares estimate b^*. This can be stated succinctly as

$$b^* = \arg\max_b L(b) = \arg\min_b S(b).$$

Hence the least squares estimator of β is the same as the maximum likelihood estimator derived under an assumption of Normality. The least squares estimator does not depend upon an assumption of Normality. In fact, no distribution need be assumed when the method of least squares is used.

To obtain the maximum likelihood estimator of σ^2, write

$$\ln L = (-n/2)\ln(2\pi) - (n/2)\ln v - S(b)/2v.$$

Differentiate this with respect to v, obtaining $\partial \ln L/\partial v = (-n/2)1/v + S(b)/2v^2$. Set this equal to 0, obtaining $(-n/2)v + S(b)/2 = 0$, and solve for v to obtain $v = S(b)/n$, the maximum likelihood estimate of σ^2. That is, the maximum likelihood estimate of σ^2 is $S(b^*)/n$.

4.3.3   A Heuristic Approach

A couple of formal estimation methods have been treated; now a heuristic method will be discussed.

4.3.3.1   Observational Equations

In terms of variables, the *statistical model* for a line through the origin is

$$Y = \beta x + \varepsilon.$$

When this is written for the n cases, as

$$Y_i = \beta x_i + \varepsilon_i, \ i = 1, 2, \ldots, n,$$

these n expressions are called the *observational equations*.

4.3.3.2   Method of Reduction of Observations

Various estimators can be obtained by working in a heuristic manner with the observational equations. They are obtained by reducing the set of n equations to a number of equations equal to the number of unknown parameters and solving them for estimates of the parameters. Among these estimators are the average of the ratios, the ratio of the averages, and the least squares estimator.

When x_i is not zero, one can divide by it to obtain

$$y_i/x_i = \beta + \varepsilon_i/x_i.$$

Solving for β gives

$$\beta = y_i/x_i - \varepsilon_i/x_i.$$

These operations can be done when each x_i is different from zero. In the example each x_i is positive because it is gallons of gasoline. One can now sum these equations and divide both sides by n to obtain

$$\beta = \frac{1}{n} \sum \frac{y_i}{x_i} - \frac{1}{n} \sum \frac{\varepsilon_i}{x_i}.$$

The term involving ε_i has mean zero and can be expected to be relatively small, because positive and negative terms will tend to cancel one another, and, in any event, the summation is divided by n, so

$$\beta \approx \frac{1}{n} \sum \frac{y_i}{x_i}.$$

So it is reasonable to take this as an estimate of β:

$$\hat{\beta} = \frac{1}{n} \sum \frac{y_i}{x_i},$$

the average-of-ratios estimate. (The *caret* $\hat{\ }$ or "hat" denotes an estimate.

Given a parameter such as β, the symbol $\hat{\beta}$, read "beta hat," denotes an estimate of β.)

An alternative method of reduction of the observational equations is simply to sum them, obtaining

$$\sum y_i = \beta \sum x_i + \sum \varepsilon_i,$$

or

$$\beta = \sum y_i / \sum x_i - \sum \varepsilon_i / \sum x_i .$$

Here the condition $\sum x_i \neq 0$ is required to be able to divide. Ignoring the term involving ε_i gives

$$\hat{\beta} = \frac{\Sigma y_i}{\Sigma x_i} = \frac{\bar{y}}{\bar{x}},$$

the ratio of averages estimate.

Still another method of reduction (Linnik 1961) is to multiply the ith equation by x_i, obtaining

$$x_i y_i = \beta x_i^2 + \varepsilon_i x_i$$

and summing to obtain

$$\sum x_i y_i = \beta \sum x_i^2 + \sum \varepsilon_i x_i,$$

or

$$\beta = \sum x_i y_i / \sum x_i^2 - \sum \varepsilon_i x_i / \sum x_i^2.$$

Here there can be no problem with the divisor $\sum x_i^2$; it is $\neq 0$ unless all $x_i = 0$. Treating the term involving ε_i as negligible, set

$$\hat{\beta} = \sum x_i y_i / \sum x_i^2,$$

which is the same as b^*, the least squares estimate.

4.3.4   Means and Variances of Estimators

4.3.4.1   Means of Estimators

The estimators considered so far are linear in the observations y_i, $i = 1, 2, \ldots, n$, that is, they are of the form $\sum_{i=1}^{n} a_i y_i$. These estimators are:

Least squares estimator: $a_i = x_i / \sum_{j=1}^{n} x_j^2$

Ratio-of-averages: $a_i = 1 / \sum_{j=1}^{n} x_j$

Average ratio: $a_i = 1/x_i$.

For now, letting $\mathcal{E}$ denote the conditional expectation given $\{x_i\}$, the expected value of a linear estimator is

$$\mathcal{E}[\sum_{i=1}^{n}(a_i Y_i)] = \sum_{i=1}^{n}\mathcal{E}[a_i Y_i]$$

$$= \sum_{i=1}^{n}a_i \mathcal{E}[Y_i]$$

$$= \sum_{i=1}^{n}a_i(\beta x_i)$$

$$= \beta(\sum_{i=1}^{n}a_i x_i),$$

that is,

$$\mathcal{E}[\sum_{i=1}^{n}(a_i y_i)] = \beta \sum_{i=1}^{n}a_i x_i. \tag{4.4}$$

4.3.4.2 Unbiasedness

An estimator is *unbiased* if its expected value is equal to the true parameter value, whatever it may be. For the three estimators under consideration, we have the following.

Least squares estimator:

$$\beta \sum a_i x_i = \beta \sum (x_i / \sum_{j=1}^{n}x_j^2)x_i = \beta \sum x_i^2 / \sum_{j=1}^{n}x_j^2) = \beta$$

Ratio-of-averages:

$$\beta \sum a_i x_i = \beta \sum (1/\sum_{j=1}^{n}x_j)x_i = \beta \sum x_i / \sum_{j=1}^{n}x_j = \beta$$

Average ratio:

$$\beta \sum a_i x_i = \beta \sum (1/x_i)x_i = \beta$$

More generally, the condition for $\sum(a_i y_i)$ to be unbiased for β is (see (4.4)) $\beta \sum a_i x_i = \beta$, for all values of β, that is, $\sum a_i x_i = 1$, or $\sum a_i x_i - 1 = 0$.

4.3.4.3 Variance of the Least Squares Estimator

The variance of a linear combination of the observations is $\mathcal{V}[\sum a_i y_i] = \sigma^2 \sum a_i^2$. For b^*, the least squares estimator, $a_i = x_i / \sum x_j^2$, so $\sum a_i^2 = \sum x_i^2/(\sum x_j^2)^2 = 1/\sum x_i^2$, and

$$\mathcal{V}[b^*] = \frac{\sigma^2}{\sum x_i^2}.$$

4.3.4.4   Nonlinear and Biased Estimators

It is not necessary to consider only linear estimators. For example, one might wish to shrink the estimate closer to some particular value β_0, using an estimator $pb + (1 - p)\beta_0$, for some proportion p between 0 and 1. The quantity p could even be taken to be a statistic, for example related to the F statistic for testing the hypothesis that $\beta = \beta_0$, where p is close to 1 if F is large and close to 0 if F is small. Such a p is $p = (1 - c/F)$, for suitable constants c. Such nonlinear estimators may be biased. Then instead of the variance, one considers the mean squared error,

$$\text{MSE}[\hat{\beta}] \;=\; \mathcal{E}[\,(\hat{\beta} - \beta)^2\,]$$

as a measure of the goodness of the estimator. It can be shown that the mean squared error equals the variance plus the square of the bias,

$$\text{MSE}[\hat{\beta}] \;=\; \mathcal{V}[\hat{\beta}] \;+\; [\,\text{Bias}[\hat{\beta}]\,]^2,$$

where $\text{Bias}[\hat{\beta}] = \mathcal{E}[\hat{\beta}] - \beta$.

4.3.5   Estimating the Error Variance

When the ε_i have a common variance σ^2, this parameter is called the *error variance*. Now consider estimating it. The model gives

$$\varepsilon_i \;=\; y_i - \beta x_i.$$

Then the *residual* $e_i = y_i - bx_i$ is unbiased for ε_i if b is any unbiased estimator for β. Because $\mathcal{E}[\varepsilon_i] = 0$, the variance is $\mathcal{V}[\varepsilon_i] = \mathcal{E}[e_i^2] = \sigma^2$. The squared residual e_i estimates ε_i. So, its square e_i^2 is a reasonable estimate of $\mathcal{E}[\varepsilon_i^2]$, which is σ^2. One can average the e_i^2 to obtain an estimate of σ^2. In averaging, the divisor is $n - 1$ rather than n. The reason for this is similar to the reason for dividing by $n - 1$ instead of n when computing a sample variance from deviations $x_i - \bar{x}$. This is related to the number of *degrees of freedom*, or d.f., associated with the vector of deviations from the mean. The sum $\sum(x_i - \bar{x}) = 0$, so the inner product of the vector of deviations and the vector of all ones is 0 and so the vector of deviations is perpendicular to the equi-angular line in n-space, and hence has only $n - 1$ d.f. In the case of fitting a line through the origin, if $n = 1$, we could fit the line to the data exactly, with no error: we simply draw the line between the origin and the one data point. Then, $e_1 = 0$. There are no d.f. for error. In general, the number of d.f. for error in fitting a line through the origin is $n - 1$. There are n residuals, but they satisfy $\sum e_i x_i = \sum(y_i - bx_i)x_i = 0$. The vector of residuals is perpendicular to the vector of x_i, so it varies in a space of dimension $n - 1$. If the values of $n - 1$ of the residuals are known, the value of the n-th is determined. Related to this fact is the fact that $\mathcal{E}[\,(\sum e_i^2\,] = (n-1)\,\sigma^2$. For,

$$
\begin{aligned}
e_i &= y_i - bx_i \\
&= y_i - \beta x_i + \beta x_i - bx_i \\
&= (y_i - \beta x_i) + (\beta - b)x_i \\
&= \varepsilon_i - (b - \beta)x_i,
\end{aligned}
$$

and for the sum of squared residuals,

$$
\begin{aligned}
\sum e_i^2 &= \sum [\varepsilon_i - (b-\beta)x_i]^2 \\
&= \sum \varepsilon_i^2 - 2(b-\beta)\sum \varepsilon_i x_i + \sum [(b-\beta)x_i)]^2.
\end{aligned}
$$

The expected value of the first term is $n\sigma^2$. The expected value of the third term is $\mathcal{E}(b-\beta)^2 \sum x_i^2 = \mathcal{V}[b]\sum x_i^2 = (\sigma^2/\sum x_i^2)\sum x_j^2 = \sigma^2$. For the second term, we note that

$$
\begin{aligned}
b &= \sum_j x_j y_j / \sum_i x_i^2 \\
&= (1/\sum_i x_i^2)\sum x_j(\beta x_j + \varepsilon_j) \\
&= \beta + \sum_j x_j \varepsilon_j / \sum x_i^2.
\end{aligned}
$$

This gives $b - \beta = \sum_j x_j \varepsilon_j / \sum_i x_i^2$, so that

$$
\begin{aligned}
\mathcal{E}[(b-\beta)\sum \varepsilon_i x_i] &= \mathcal{E}[(\sum_j \varepsilon_j x_j)(\sum_k \varepsilon_k x_k)]/\sum x_i^2 \\
&= \sum_j \sum_k x_j x_k \mathcal{E}(\varepsilon_j \varepsilon_k)/\sum_i x_i^2 \\
&= \sum_j \sum_k x_j x_k \delta_{jk}\sigma^2 / \sum_i x_i^2 \\
&= \sigma^2 \sum_j x_j^2 / \sum_i x_i^2 \\
&= \sigma^2.
\end{aligned}
$$

Collecting terms gives the expectation as $(n - 2 + 1)\sigma^2 = (n-1)\sigma^2$. So, $\mathcal{E}[\,(\sum e_i^2\,] = (n-1)\sigma^2$, and $\mathcal{E}[\sum e_i^2/(n-1)] = \sigma^2$; that is, $\sum e_i^2/(n-1)$ is an unbiased estimator for σ^2. This statistic is denoted by s^2 and is called the *residual mean square*: $s^2 = \sum e_i^2/(n-1)$.

4.3.5.1   Computational Formulas

The sum of squared residuals or sum of squared "errors" is denoted by SSE. We have

$$
\begin{aligned}
\text{SSE} \; &= \; \sum e_i^2 \\
&= \; \sum (y_i - bx_i)^2 \\
&= \; \sum y_i^2 - 2b \sum x_i y_i + b^2 \sum x_i^2 \\
&= \; \sum y_i^2 - 2b(b \sum x_i^2) + b^2 \sum x_i^2 \\
&= \; \sum y_i^2 - b^2 \sum x_i^2 ,
\end{aligned}
$$

which may also be written as $\sum y_i^2 - (\sum x_i y_i)^2 / \sum x_i^2$.

4.3.5.2   Decomposition of Sum of Squares

The model here is

$$
Y_i \; = \; \mathcal{E}[Y_i \,|\, x_i] \; + \; (Y_i - \mathcal{E}[Y_i \,|\, x_i]) = \beta x_i + \varepsilon_i,
$$

where the "error" is $\varepsilon_i = Y_i - \mathcal{E}[Y_i | x_i]$. This is a *decomposition of the random variable* into its conditional mean plus error, the error being the difference from the conditional mean,

$$
\text{RESPONSE} \; = \; \text{CONDITIONAL EXPECTATION} + \text{ERROR}.
$$

Such models are called *additive* because the error term is added. Analogous to this decomposition is the *decomposition of the observations,*

$$
y_i = \hat{y}_i + (y_i - \hat{y}_i) = b^* x_i + (y_i - b^* x_i) = b^* x_i + e_i,
$$

or

$$
\text{OBSERVATION} \; = \; \text{FITTED VALUE} + \text{RESIDUAL},
$$

a decomposition of the observation into its fitted value $\hat{y}_i = b^* x_i$ plus a residual $e_i = y_i - b^* x_i$, that is a difference from the fitted value. The sum of squares is

$$
\sum y_i^2 = b^{*2} \sum x_i^2 - 2b^* \sum x_i e_i + \sum e_i^2 = b^{*2} \sum x_i^2 + \sum e_i^2, \qquad (4.5)
$$

because $\sum x_i e_i = \sum x_i (y_i - b^* x_i) = \sum x_i y_i - b^* \sum x_i^2 = \sum x_i y_i - (\sum x_i y_i / \sum x_i^2) \sum x_i^2 = \sum x_i y_i - \sum x_i y_i = 0$. This expression is a *decomposition of the sum of squares* into a part due to regression and a part due to residuals,

$$
\text{SST} \; = \; \text{SSR} \; + \; \text{SSE},
$$

where SST is the total sum of squares $\sum y_i^2$, $\text{SSR} = b^{*2} \sum x_i^2 = b^* \sum x_i y_i$, and $\text{SSE} = \sum e_i^2$.

4.4   Inference Concerning the Slope

Inference about the parameter β is considered next.

4.4.1   Testing a Hypothesis Concerning the Slope

First consider the null hypothesis $H_0 : \beta = 0$. The ratio $t = b^*/SE[b^*]$ is the estimate b^*, expressed in standard deviation units. The standard deviation of b^* is $SD[b^*] = \sqrt{V[b^*]} = \sqrt{\sigma^2/\sum x_i^2} = \sigma/\sqrt{\sum x_i^2}$. The term "standard error" is used here to mean an estimate of the standard deviation of a statistic. The standard error of b^* is $SE[b^*] = s/\sqrt{\sum x_i^2}$. It is reasonable to reject the null hypothesis when this ratio is large, say, to accept the null hypothesis when $-2 < b^*/SE(b^*) < +2$. Of course, instead of 2 we should have a value that is an appropriate percentile of the null distribution of this test statistic.

It can be shown that, assuming Normality, the distribution of the ratio of the estimate to its standard error, $b^*/SE[b^*]$, is the Student's t distribution with $n - 1$ d.f. Denote the upper p-th percentile of this distribution by $t_{n-1,p}$. Then the acceptance region of the two-tailed level-α test is $-t_{n-1,\alpha/2} < b^*/SE[b^*] < t_{n-1,\alpha/2}$. When n is large and $\alpha = .05$, $t_{n-1,\alpha/2} \approx 1.96$, or about 2.

Somewhat more generally, consider testing $H_0 : \beta = \beta_0$, where β_0 is a specified value of β. (In the CAPM, to be discussed below, one might want to test $\beta = 1$.) The t test statistic for this problem is $(b^* - \beta_0)/SE[b^*]$.

The t test is generally robust (still gives reasonable results) against mild violations of the assumptions of Normality and *homoscedasticity*, that is constant variance. (Non-constant variance is called *heteroscedasticity*.)

4.4.2   Confidence Interval

When the errors are Normally distributed, uncorrelated and have the same variance, $(b^* - \beta_0)/SE[b^*]$ has a t distribution with $n - 1$ d.f. when $\beta = \beta_0$. It follows then that a $100(1 - \alpha)\%$ confidence interval for β is

$$(b^* - \text{m.e.}, \ b^* + \text{m.e.}),$$

where m.e. is the margin of error, $t_{n-1,\alpha/2} SE[b^*]$, . where $t_{m,p}$ denotes the upper $100p$-th percentage point of the t distribution with m degrees of freedom.

Remember that a confidence interval can be considered to be a range of plausible (reasonable) values for the parameter, given the data. In particular, if a value β_0 is in the interval, then this value is plausible.

4.5   Testing Equality of Slopes of Two Lines through the Origin

Suppose that Jones made runs 1 through 10 using regular gasoline and runs 11 through 14 using premium gasoline. He wants to compare the miles per gallon of the two grades of gasoline. He will test $H_0 : \beta_1 = \beta_2$ against one-sided alternatives $H_a : \beta_2 > \beta_1$, where the subscript 1 refers to regular and 2 to premium. The test statistic is $(b_2 - b_1)/\mathrm{SE}[b_1 - b_2]$. One then sees where the value of this test statistics falls in the t distribution with $n_1 + n_2 - 2 = 10 + 4 - 2 = 12$ d.f.

We give further details in general terms and leave the completion of this example as an exercise. Write the observations in the two groups as $(x_{gi}, y_{gi}), g = 1, 2, \; i = 1, 2, \ldots, n_g$. Then the variance of the difference between the two statistics, which are uncorrelated (in fact, statistically independent), is

$$
\begin{aligned}
\mathcal{V}[b_1 - b_2] &= \mathcal{V}[b_1] + \mathcal{V}[b_2] \\
&= \sigma^2 / \sum x_{1i}^2 + \sigma^2 / \sum x_{2i}^2 \\
&= \sigma^2 (1/\sum x_{1i}^2 + 1/\sum x_{2i}^2).
\end{aligned}
$$

The estimates of the slopes are, for $g = 1, 2, \; b_g = \sum_i x_{gi} y_{gi} / \sum_i x_{gi}^2$. The residual sum of squares is

$$
S = \sum_g \sum_i (y_{gi} - b_g x_{gi})^2 = \sum_g \sum_i y_{gi}^2 - b_1^2 \sum x_{1i}^2 - b_2^2 \sum x_{2i}^2,
$$

with $n_1 + n_2 - 2$ d.f. The estimate of σ^2 is $s^2 = S/(n_1 + n_2 - 2)$. The estimate of the variance of the difference between b_1 and b_2 is $s^2(1/\sum x_{1i}^2 + 1/\sum x_{2i}^2)$. The standard error of the difference $\mathrm{SE}[b_1 - b_2]$ is

$$
\mathrm{SE}[b_1 - b_2] = s \sqrt{1/\sum x_{1i}^2 + 1/\sum x_{2i}^2}.
$$

The test statistic is $t = (b_1 - b_2)/\mathrm{SE}[b_1 - b_2]$.

In the example, the variance of the difference is $\sigma^2 (1/\sum_{i=1}^{10} x_i^2 + 1/\sum_{i=11}^{14} x_i^2)$, estimated by $s^2(1 \sum_{i=1}^{10} x_i^2 + 1/\sum_{i=11}^{14} x_i^2)$. The standard error $\mathrm{SE}[b_1 - b_2]$ is the square root of this. For this two-sample problem, s^2 is computed as $[\sum_{i=1}^{10} (y_i - b_1 x_i)^2 + \sum_{i=11}^{14} (y_i - b_2 x_i)^2]/12$. The divisor is $n - 2 = 14 - 2 = 12$. Further details for carrying out the test are left as an exercise.

4.6 Linear Parametric Functions

What may be more to the point is whether

Cost per mile for premium < Cost per mile for regular.

Assume that regular gasoline costs c_{reg} per gallon and premium costs c_{prem} ($c_{prem} > c_{reg}$) per gallon. Because

$$\text{Cost per mile} = \frac{\text{Cost per gallon}}{\text{Miles per gallon}},$$

$$\text{Miles per dollar} = \frac{\text{Miles per gallon}}{\text{Dollars per gallon}}.$$

For premium this is β_2/c_{prem} and for regular β_1/c_{reg}. We want to make inferences about the difference, miles per dollar for regular minus miles per dollar for premium, that is, $\beta_1/c_{reg} - \beta_2/c_{prem}$. If this is positive, regular is more cost efficient. We can test

$$H_0 : \beta_1/c_{reg} - \beta_2/c_{prem} = 0 \quad \text{against} \quad H_a : \beta_1/c_{reg} - \beta_2/c_{prem} \neq 0$$

and make a confidence interval for $\beta_1/c_{reg} - \beta_2/c_{prem}$.

Such a linear combination of parameters is called a *linear parametric function*. We denote such a quantity by ψ. Here it is of the form $\psi = c_1\beta_1 + c_2\beta_2$. It is also called a linear *contrast*. The constants c_1, c_2 are called the *contrast coefficients*. The estimate of ψ is $\hat{\psi} = c_1b_1 - c_2b_2$. The variance of the estimate is

$$\mathcal{V}[\hat{\psi}] = c_1^2 \mathcal{V}[b_1] + c_2^2 \mathcal{V}[b_2] = \sigma^2 \left(c_1^2 / \sum x_{1i}^2 + c_2^2 / \sum x_{2i}^2 \right).$$

The estimate of this is $s^2 \left(c_1^2 / \sum x_{1i}^2 + c_2^2 / \sum x_{2i}^2 \right)$. The standard error of the estimate of the contrast is the square root of this,

$$\mathrm{SE}[\hat{\psi}] = s \sqrt{c_1^2 / \sum x_{1i}^2 + c_2^2 / \sum x_{2i}^2}.$$

The test statistic is $t = \hat{\psi}/\mathrm{SE}[\hat{\psi}]$. Assuming approximate Normality, this has a t distribution and the $100(1-\alpha)\%$ confidence interval is $(\hat{\psi} - \text{m.e.}, \hat{\psi} + \text{m.e.})$, where the margin of error is m.e. $= t^* \, \mathrm{SE}[\hat{\psi}]$ and $t^* = t_{n-2,\alpha/2}$, the upper $\alpha/2$ percentage point of the t distribution with $n-2$ d.f., where $n = n_1 + n_2$.

The reader is asked to complete the example as an exercise.

4.7 Variances Dependent upon X*

Consider a model in which $x_i > 0$ and $\mathcal{V}[\varepsilon_i] = \sigma^2 x_i^p$, where p could be any real number, but the values $p = 0, 1$, and 2 are of special interest. The case $p = 0$ gives the homoscedastic case.

We work on the principle that the least squares criterion is appropriate for n cases that are uncorrelated and of equal variance. We shall see in later chapters that an appropriate criterion can be obtained by transforming to uncorrelated variables of equal variance, writing the least squares criterion in terms of them, and then transforming back to the original variables to obtain a criterion in terms of them. In the present model this merely means dividing by the variance of the i-th case. We take as the estimate of the parameter β the value b^* of b that minimizes the least squares criterion

$$
\begin{aligned}
S(b) &= \sum (y_i - b\,x_i)^2 / \mathcal{V}[Y_i] \\
&= \sum (y_i - bx_i)^2 / \sigma^2 x_i^p \\
&= \sum [(y_i - bx_i)/x_i^{p/2}]^2 / \sigma^2 .
\end{aligned}
$$

The estimate can be written as

$$
b^* = \arg \min_b S(b).
$$

Thus, the problem is to minimize $\sum [(y_i - bx_i)/x_i^{p/2}]^2$ with respect to b. This equals $\sum (v_i - bu_i)^2$, where $u_i = x_i/x_i^{p/2} = x_i^{1-p/2}$ and $v_i = y_i/y_i^{p/2} = y_i^{1-p/2}$. In terms of U and V, the least squares estimate is $b = \sum u_i v_i / \sum u_i^2$; in terms of X and Y, this is $b_p = \sum y_i x_i^{1-p} / \sum x_i^{2-p}$. The value $p = 0$ gives the ordinary least squares estimate; the value $p = 1$, the ratio of averages; the value $p = 2$, the average of ratios.

When one does not wish to assume a particular value for p, it also can be considered to be a parameter, say λ, and estimated from the data. This can be done by simple inspection of a plot of the squared residuals against X, possibly supplemented with the computation of some correlations, or, in the Normal case, a maximum likelihood analysis. Next we elaborate these methods.

Let b be any unbiased estimate of β, perhaps the ordinary least squares estimate, b^*, and compute the residuals $e_i = y_i - bx_i$. Plot their squares against x_i, because the squared residual estimates the variance at that point. If the plot looks like a random swarm of points, $p = 0$ is perhaps justified. If the plot increases linearly, $p = 1$ is suggested. If the plot curves, $p = 2$ is suggested. One might also supplement this analysis by correlating the squared residuals with X and with X^2.

If it is assumed that the ε_i have a distribution in some particular family, then p as well as the other parameters can be estimated by the method of maximum likelihood. If one assumes tha the errors are distributed according to a Normal distribution, then the likelihood is

$$L(p,b,v) = \prod_{i=1}^{n}(2\pi v x_i^p)^{-1/2}\exp[-(y_i - bx_i)^2/(2vx_i^p]$$

$$= (2\pi v)^{-n/2}(\prod_{i=1}^{n}x_i^p)^{-1/2}\exp[-\sum(y_i - bx_i)^2/(2vx_i^p],$$

where we represent the parameters λ, β, and σ^2 by variables p, b, and v. It is required to find the values that achieve $\max_{p,b,v} L(p,b,v)$. The same values that maximize L also maximize $\ln L$; that is, maximizing $L(p,b,v)$ is equivalent to maximizing $K(p,B,v)$, where $K = \ln L$. Due to the form of the function involved, it is convenient to do this in the order $\max_p[\max_v(\max_b K)]$. That is, we maximize over b for given p, v, then maximize the result with respect to v for given p. Finally, we maximize this result over p in the specified set, usually $p = 0, 1, 2$.

For any fixed values p and v, we maximize over b and obtain the estimate b_p Then $\max_b K(p,B,v) = -n/2\ln(2\pi v) - (p/2)\ln\sum x_i - \sum(y_i - b_px_i)^2/2vx_i^p = -n/2\ln(2\pi v) - (p/2)\sum\ln x_i - S_p/2v$, where

$$S_p = \sum[(y_i - b_px_i)^2/x_i^p]. \tag{4.6}$$

To maximize over v, we compute $\partial[-n/2\ln(2\pi v) - (p/2)\sum\ln x_i - S_p/2v]/\partial v = -n/2v - S_p/2v^2 = 0, nv = S_p, v = S_p/n$. Thus $\max K_{b,v} = -(n/2)\ln(2\pi S_p/n) - (p/2)\sum\ln x_i - n/2$. The maximum likelihood estimate of the power parameter λ is the value of p that maximizes this, that is, which minimizes $(n/2)\ln(2\pi S_p/n) + (p/2)\sum\ln x_i + n/2$, or, equivalently, which minimizes $n\ln S_p + p\sum\ln x_i$, where S_p is given in Equatioin 4.6. Note that the maximum likelihood estimate of the power parameter λ represented by the variable p depends not only upon the residual sum of squares, S_p, but also upon the term $p\sum\ln x_i$ in the Normal case. When one is unable to specify a family of distributions, one could argue that S_p is at least somewhat indicative of the value of p.

4.8 A Financial Application: CAPM and "Beta"

At this point we return to financial investments analysis with an example of a "market model." "Market models" describe the rates of return of assets in terms of the rate of return and other characteristics of the market as a whole. Here, the term *asset* refers to shares of a stock, a mutual fund, or a portfolio (set of stocks).

Price series and rates of return (RORs) were discussed in the preceding chapter. The "raw" (original) data are the prices P_{at}, $a = 1, 2, \ldots, m$ assets, $t = 1, 2, \ldots, n$. The RORs $R_{at} = (P_{at} - P_{a,t-1})/P_{a,t-1}$, which are *derived* data, are to be the object of analysis here.

Let R_{mt} denote the ROR of the market as a whole, as indicated, for example, by the ROR of the S&P500 composite stock index. (The S&P 500 index has several ticker symbols, including: $\hat{\text{G}}$SPC[2] and $SPX.) The discussion here is in terms of R, the ordinary ROR, but continuous ROR r could be used as well. Let R_{ft} denote the risk-free ROR, as represented for example by the ROR of one-month or three-month treasury bills. One can realize an ROR R_{ft} without investing in a risky asset such as shares of stock.

4.8.1   CAPM

The quantity $R_{at} - R_{ft}$ is called the *excess* ROR of asset a at time t. The quantity $R_{mt} - R_{ft}$ is the excess ROR of the market at time t. The excess ROR is the ROR associated with the decision to hold the risky asset i rather than a risk-free instrument. The term "excess" here does not mean "extra" or "abnormal." It simply means the amount above the the the risk-free ROR. Later, in discussing portfolio analysis, it will be seen that optimization in terms of several risky assets and a risk-free asset comes out in terms of excess returns.

The *Capital Asset Pricing Model*, or CAPM, asserts that the expected value of the excess ROR of asset i is equal to a coefficient β_i times the excess ROR of the market:

$$R_{at} - R_{ft} = \beta_a (R_{mt} - R_{ft}) + \varepsilon_{it}.$$

For asset i this is of the form

$$y_{at} = \beta_a x_t + \varepsilon_{at},$$

the form of model discussed above, with $y_{at} = R_{at} - R_{ft}$ and $x_t = (R_{mt} - R_{ft})$.

In general, the units of β are units of y divided by units of x. If y is weight in pounds and x is height in inches, then the units of β in the regression of y on x are pounds per inch. The meaning of β is that if x increases by one x-unit, then the expected value of y increases by β y-units.

In the CAPM, both y and x are rates of return. In particular, they have the same units. When y and x have the same units, then β is a pure number, in the sense of being dimensionless (unitless)—the units cancel out. The meaning of β in the CAPM is that if x increases by δx, then the (conditional) expected value of Y increases by $\beta \, \delta x$. If $\beta = 1$, an increase of 1% in x is associated with a 1% one percent increase in the conditional expected value of Y. If $\beta = 2$, an increase of 1% in x is associated with a 2% increase in the conditional expected value of Y. If $\beta = 1/2$, an increase of 1% in x is associated with a 0.5%a increase in the conditional expected value of Y.

For one asset and the market, the dataset can be considered to be in the form for simple linear regression,

$$\{(x_i, y_i), i = 1, 2, \ldots, n\}.$$

The usual and customary size of a dataset for this purpose is five years of monthly data ($n = 60$ months).

4.8.2 "Beta"

Usually, β_a will be positive, because most assets a are positively correlated with the market as a whole. If $\beta_a > 1$, asset a is expected to increase by more than 1% if the market increases by 1%. If $0 < \beta_a < 1$, asset a is expected to increase by less than 1% if the market increases by 1%. Of course, if the market decreases by 1%, an asset a with $0 < \beta_a < 1$ is expected to decrease by less than 1%. Such an asset, with a beta of less than 1, is called "conservative." An asset with a beta of more than 1 is called "risky." But sometimes a dividing point more like 1.5 might be used instead of 1.0.

4.9 Slope and Intercept

Even when $x = 0$ implies $y = 0$, a model with both a constant and a slope may be more descriptive of the data. Such is the case, for example, with weight and height, where the fitted weight w in terms of height h might be something like $\hat{w} = 4.4\,h - 137$ lb.

4.9.1 Model with Slope and Intercept

So, often a model with both a slope β and an intercept (constant) α is used. The *statistical model* is

$$Y = \alpha + \beta X + \varepsilon.$$

The conditional mean $\mathcal{E}[Y \mid x]$ is the regression function $\eta(x) = \alpha + \beta x$. (The letter η, Greek y, denotes a mathematical expectation of the r.v. Y.)

The data for simple linear regression take the form of (x, y) pairs for n cases or instances of observation, $(x_i, y_i), i = 1, 2, \ldots, n$. The *observational model* is the statistical model, stated in terms of cases. This gives the *observational equations*, which here are

$$Y_i = \alpha + \beta x_i + \varepsilon_i, i = 1, 2, \ldots, n.$$

The conditional mean of Y_i given x_i is $\alpha + \beta x_i$. The variables ε_i are differences from the conditional mean and hence have mean 0.

Note that the presence of the additional term α changes the meaning of and estimate of the slope parameter β. (The β in $\mathcal{E}[Y \mid x] = \alpha + \beta x$ is not the same as the β in $\mathcal{E}[Y\,|\,x] \;=\; \beta\,x$. The value and meaning of any such parameter is dependent upon what other terms are in the model.)

The least squares criterion evaluated for trial values a and b is

$$S(a,b) \;=\; \sum_{i=1}^{n} [y_i - (a + bx_i)]^2.$$

Minimizing by computing partial derivatives with respect to a and b and setting them equal to zero (or otherwise) leads to two simultaneous linear equations, the solutions of which are the least squares estimates a^*, b^*. This can be written as

$$(a^*,\, b^*) \;=\; \arg\min_{a,b} S(a,\,b).$$

These least squares estimates are

$$b^* = [\sum_{i=1}^{n} x_i\,y_i \;-\; (\sum_{i=1}^{n} x_i)(\sum_{i=1}^{n} y_i)/n] \,/\, [\sum_{i=1}^{n} x_i^2 - (\sum_{i=1}^{n} x_i)^2/n]$$

and

$$a^* = [\sum_{i=1}^{n} y_i \;-\; b \sum_{i=1}^{n} x_i]/n.$$

These can be written as

$$b^* \;=\; \sum_{i=1}^{n} (x_i - \bar{x})(y_i - \bar{y}) \,/\, \sum_{i=1}^{n} (x_i - \bar{x})^2$$

and

$$a^* \;=\; \bar{y} - b\,\bar{x}.$$

(Alternatively, one can write $y_i = \gamma + \beta(x_i - \bar{x}) + \varepsilon_i$, where $\gamma = \alpha + \beta\bar{x}$, note that the estimate of γ is simply $\bar{y}$, and proceed from there to get the same a^*, b^*.)

4.9.2 CAPM with Differential Return

In the context of the CAPM, an intercept α is often used to represent the systematic part of ROR that is not explained by the ROR of the market. This part is called the *differential* or *abnormal* ROR. See Jensen (1968). Seeking investments that yield positive differential returns is sometimes called "chasing alpha." The model is $R_{at} - R_{ft} \;=\; \alpha_a + \beta_a\,(R_{mt} - R_{ft})$, $t = 1, 2, \ldots, n$, for assets a taken one at a time.

4.10 Appendix 4A: Optimality of the Least Squares Estimator

Next the optimality of the least squares estimator is illustrated in the case of estimation of a single regression parameter.

The Cauchy–Schwarz inequality. Let $a_i, x_i, i = 1, 2, \ldots, n$, be any $2n$ numbers. Then

$$\left(\sum a_i x_i \right)^2 \le \left(\sum a_i^2 \right)\left(\sum x_i^2 \right).$$

Proof: For any number b, $0 \le \sum (a_i - b x_i)^2 = \sum a_i^2 - 2b \sum a_i x_i + b^2 \sum x_i^2 = f(b)$, say. Taking $b = \sum a_i x_i / \sum x_i^2$ gives

$$0 \le f\left(\sum a_i x_i / \sum x_i^2\right) = \sum a_i^2 - \left(\sum a_i x_i\right)^2 / \sum x_i^2.$$

Multiplying through by $\sum x_i^2$ gives the result.

Fact. Let $b = \sum a_i y_i$ by any unbiased linear estimator for β. Then $\mathcal{V}[b] \ge \mathcal{V}[b] = \sigma^2 / \sum x_i^2$. That is, the least squares estimator b^* has minimum variance in the class of all unbiased linear estimators.

Proof: By the assumption of unbiasedness, $\beta = \mathcal{E}[b] = \mathcal{E}[(\sum a_i Y_i)] = \sum a_i \mathcal{E}[Y_i] = \sum a_i \beta x_i = \beta \sum a_i x_i$, for all values of β, which implies that $\sum a_i x_i = 1$. Now, $\mathcal{V}[b] = \mathcal{V}[\sum a_i y_i] = \sum a_i^2 \mathcal{V}[Y_i] = \sigma^2 \sum a_i^2$. Thus, it suffices to prove that $\sum a_i^2 \ge 1 / \sum x_i^2$. But, by the Cauchy–Schwarz inequality, $\sum a_i^2 \sum x_i^2 \ge (\sum a_i x_i)^2 = 1^2 = 1$, so $\sum a_i^2 \ge 1 / \sum x_i^2$, as was to be shown.

Remark. The Cauchy–Schwarz inequality shows that the size of the correlation coefficient is less than or equal to one, that is, $|r| \le 1$, or $-1 \le r \le +1$. To see this, given $(u_i, v_i), i = 1, 2, \ldots, n$, take $a_i = u_i - \bar{u}$ and $x_i = v_i - \bar{v}$.

These methods and results generalize to the case of slope and intercept and to the case of regression on several variables (multiple regression).

4.11 Summary

The conditional expected value of a variable Y given that another variable X has a particular value, say x, is the mean of the distribution of Y for instances in which $X = x$. This conditional expectation is denoted by $\mathcal{E}[Y \mid X = x]$.

The *simple linear regression* model describes the conditional expected value of a response (dependent) variable Y as a linear function of an explanatory (independent) variable x. The word "simple" refers to the fact that

Y is being described in terms of a single explanatory variable x. In "multiple" regression, the conditional expectation of Y is modeled in terms of several explanatory variables.

The parameters of a simple linear regression model are the slope β, the intercept α, and the variance σ^2. This is the variance of the conditional distribution of Y given x.

The *statistical model* is $Y = \alpha + \beta x + \varepsilon$. The *observational equations* are $y_i = \alpha + \beta x_i + \varepsilon_i$, $i = 1, 2, \ldots, n$. The variance of ε_i is σ^2.

The least squares criterion for evaluating a, b as estimates of α, β is $S(a, b) = \sum_{i=1}^{n} [y_i - (a + b x_i)]^2$. The principle of least squares says to choose as the estimates those values a^* for a and b^* for b that minimize the least squares criterion.

Regression through the origin refers to a model where $\mathcal{E}[Y \,|\, x] = \beta x$.

In the CAPM (Capital Asset Pricing Model), the variable Y is the excess ROR of any given asset, and x is the excess ROR of the market.

4.12   Chapter Exercises

4.12.1   Applied Exercises

4.1 Jones' mpg data are in Table 4.1.

  a. Compute the least squares estimate of the mpg.

  b. Compute the "ratio of means" estimate.

  c. Compute the "mean of the ratios" estimate.

4.2 Suppose there was an error in recording Run 9. The values should be (98, 12.8) instead of (96, 7.5). The corrected dataset is in Table 4.4.

  a. Compute the least squares estimate of the mpg.

  b. Compute the "ratio of means" estimate.

  c. Compute the "mean of the ratios" estimate.

4.3 Suppose that Jones made runs 1 through 10 using regular gasoline and runs 11 through 14 using premium.

  a. Compute the least squares estimates b_1 and b_2 of the mpg obtained using regular and premium.

  b. Test the hypothesis of equality of miles per gallon, against the appropriate one-sided alternatives.

TABLE 4.4
Gasoline Mileage Data

Run	Miles	Gallons	Run	Miles	Gallons
1	62	4.6	8	108	8.0
2	49	5.7	9	98	12.8
3	73	4.3	10	61	4.7
4	63	6.1	11	165	10.8
5	108	9.3	12	148	8.9
6	135	9.9	13	197	13.1
7	60	4.9	14	185	12.7

4.4 The car dealer told Jones he should expect to get at least 15 mpg. (It is an old-model, used car.) Jones bought the car. The data from a number of fill-ups are in Table 4.4. The least squares estimate is only 13.95 mpg, 1.05 mpg lower than the claimed 15 mpg. Is 13.95 mpg significantly smaller than 15.0 mpg? Test $H_0 : \beta = 15$ against alternatives $H_a : \beta < 15$. ANSWER: $t = 2.08$, 13 d.f., one-tailed $p = .029$ (at the .05 level, for example, reject H_0).

4.5 Compare the mpg of regular and premium gasoline (runs 1–10 and 11–14) by testing the relevant hypothesis. (Using methods to be discussed in later chapters, this can be done by introducing a dummy variable that is 0 for regular runs and 1 for premium runs.)

4.6 Cost comparison. Refer to the data in Table 4.1. Compare the dollars per gallon of regular (runs 1–10) and premium (runs 11–14), as outlined in the text and using the costs per gallon given there. Make a t test at the .05 level and a 95% confidence interval for the contrast involved.

4.7 The Sarabee Foods Company buys beef for some of its items. The fat must be trimmed off. Table 4.5 shows total weight and weight of fat for sixteen purchases of Grade A and fourteen purchases of Grade B beef, sorted within grades by total weight.

a. Compute the least squares estimate of the proportion of the total weight that is fat for each grade.

b. Compute the "ratio of means" estimate.

c. Compute the "mean of the ratios" estimate.

4.12.2 Mathematical Exercises

4.8 Given: two points P_1 with coordinates $(x_1, y_1) = (1, 3)$ and P_2 with coordinates $(x_1, y_2) = (3, 5)$. What is the slope of the line through them?

TABLE 4.5

Data for Beef Purchases (weight in ounces).

Grade A (n=16)		Grade B (n=14)	
Total Weight	Fat	Total Weight	Fat
12	0.6	10	1.0
16	0.8	12	1.1
17	0.8	16	1.2
18	0.9	17	1.3
19	0.9	18	1.3
20	1.0	19	1.5
20	1.1	19	1.5
21	1.2	20	1.7
22	1.3	22	1.8
24	1.4	26	2.4
26	1.5	28	2.4
28	1.5	28	2.4
30	1.5	31	2.6
31	1.5	32	2.7
32	1.5		
33	1.5		

4.9 (continuation) What is the point half-way between the two given points?

4.10 (continuation) What is the equation of the line perpendicular to this line and through the point (x_1, y_1)? *Hint:* The slope is the negative reciprocal of that of the given line.

4.11 Given: two points P_1 with coordinates $(x_1, y_1) = (0, 0)$ and P_2 with coordinates (x_2, y_2). What is the slope of the line through them?

4.12 (continuation) What is the point half-way between the two given points?

4.13 (continuation) What is the equation of the line perpendicular to this line and through the point (x_1, y_1)? *Hint:* The slope is the negative reciprocal of that of the given line.

4.14 Model with just a mean. Specialize the regression function $\mathcal{E}[Y \mid x] = \alpha + \beta x$ to the situation in which $\mathcal{E}[Y \mid x] = \mu$, for all x. *Answer:* $\beta = 0$ and $\alpha = \mu$ (or $\alpha = 0$, $\beta = \mu$ and $x = 1$ for every case).

4.15 The sample mean is a least squares estimate. Show that the sample mean is the least squares estimate of μ in the model $Y_i = \mu + \varepsilon_i$.

4.13 Bibliography

Graybill, Franklin A. (2000). *Theory and Application of the Linear Model. Duxury Classic edition.* Duxbury Resource Center, Pacific Grove, CA. (Original copyright 1976.)

Jensen, Michael C. (1968). The performance of mutual funds in the period 1945–1964. *Journal of Finance,* **23,** 389–416.

Linnik, Yu. V. (1961). *Method of Least Squares and Principles of the Theory of Observations.* Pergamon Press, New York, Oxford, London, Paris.

Lintner, John (1965). The valuation of risk assets and the selection of risky investments in stock portfolios and capital budgets. *Review of Economics and Statistics,* **47,** 13–37.

Sharpe, William F. (1964). Capital asset prices: A theory of market equilibrium under conditions of risk. *Journal of Finance,* **19,** 425–442.

Treynor, Jack L. (1961). Market Value, Time, and Risk. Unpublished manuscript.

Treynor, Jack L. (1962). Toward a Theory of Market Value of Risky Assets. Unpublished manuscript. (Later version: Treynor 1999.)

Treynor, Jack L. (1999). Toward a theory of market value of risky assets, pp. 15–22 in Robert A. Korjczyk (Ed.), *Asset Pricing and Portfolio Performance: Models, Strategy and Performance Metrics.* Risk Books, London.

4.14 Further Reading

For details on the optimality of least squares estimates in general, the reader may consult texts on linear statistical models in general, such as that by Graybill (2000).

5

Multiple Regression and Market Models

CONTENTS

5.1   Multiple Regression Models

Multiple regression models describe the conditional expectation of Y as a function of the values of p variables $X_1, X_2, \ldots, X_p$. The variable Y is called the *response variable* or *dependent variable*. The variables $X_1, X_2, \ldots, X_p$ are called *explanatory variables, independent variables,* or *predictors.*

5.1.1   Regression Function

The conditional expectation of Y, given values $x_1, x_2, \ldots, x_p$ of $X_1, X_2, \ldots, X_p$, denoted by $\mathcal{E}[\,Y \,|\, x_1, x_2, \ldots, x_p\,]$, is called the *regression function.* This func-tion is the mean of the conditional distribution of Y, given that $X_1 = x_1, X_2 = x_2, \ldots, X_p = x_p$. That is, it is the mean of Y for cases in which $X_1 = x_1, X_2 = x_2, \ldots,$ and $X_p = x_p$. Sometimes we write this function as $\eta(x_1, x_2, \ldots, x_p)$. (The notation η, Greek letter y, is used for the mathematical expectation of the r.v. Y.)

A frequently used model is that in which the regression function is linear in $x_1, x_2, \ldots, x_p$, that is, it is of the form

$$\eta(x_1, x_2, \ldots, x_p) \;=\; \alpha + \beta_1\, x_1 + \beta_2\, x_2 + \cdots + \beta_p\, x_p,$$

where $\alpha, \beta_1, \beta_2, \ldots, \beta_p$ are unknown parameters, to be estimated. The *statis-tical model* is

$$Y \;=\; \alpha + \beta_1\, X_1 + \beta_2\, X_2 + \cdots + \beta_p\, X_p + \varepsilon.$$

The *observational model* is the statistical model, stated in terms of cases. This gives the *observational equations,* which here are

$$Y_i \;=\; \alpha + \beta_1 x_{1i} + \beta_2\, x_{2i} + \cdots + \beta_p x_{pi} + \varepsilon_i, \; i = 1, 2, \ldots, n.$$

The conditional mean of Y_i, given $x_{1i}, x_{2i}, \ldots, x_{p;i}$ is $\alpha + \beta_1 x_{1i} + \beta_2 x_{2i} + \cdots + \beta_p\, x_{pi}$. The variables ε_i are differences from the conditional mean and hence have mean 0.

Estimation is discussed next.

5.1.2   Method of Least Squares

The dataset for multiple regression is of the form

$$\{\, (y_i, \; x_{1i}, \; x_{2i}, \; \ldots, \; x_{pi}\,), \; i = 1, 2, \ldots, n\}.$$

Given trial values $a, b_1, b_2, \ldots, b_p$, the corresponding fitted values of Y are

$$\hat{y}_i(a, b_1, b_2, \ldots, b_p) \;=\; a + b_1 x_{1i} + b_2 x_{2i} + \cdots + b_p x_{pi}.$$

The *method of least squares* consists of minimizing the distance between the fitted values

$$(\hat{y}_1(a, b_1, b_2, \ldots, b_p) \quad \hat{y}_2(a, b_1, b_2, \ldots, b_p) \quad \ldots \quad \hat{y}_n(a, b_1, b_2, \ldots, b_p))$$

and the actual values

$$(y_1 \quad y_2 \quad ; \ldots \quad y_n).$$

Minimizing this distance D is equivalent to minimizing its square, D^2; that is,

$$\min_{a, b_1, b_2, \ldots, b_p} D(a, b_1, b_2, \ldots, b_p) = \sqrt{\min_{a, b_1, b_2, \ldots, b_p} D^2(a, b_1, b_2, \ldots, b_p)}$$

$$= \sqrt{\min_{a, b_1, b_2, \ldots, b_p} S(a, b_1, b_2, \ldots, b_p)},$$

so it is equivalent to minimizing S. The function S is the *least squares criterion,* the sum of squared deviations between the actual values and fitted values corresponding to $a, b_1, b_2, \ldots, b_p$:

$$S(a, b_1, b_2, \ldots, b_p) = \sum_{i=1}^{n} [(y_i - \hat{y}_i(a, b_1, b_2, \ldots, b_p)]^2$$

$$= \sum_{i=1}^{n} [y_i - (a + b_1 x_{1i} + b_2 x_{2i} + \cdots + b_p x_{pi})]^2.$$

Taking partial derivatives and setting them equal to zero gives a set of $p + 1$ simultaneous linear equations for $a, b_1, b_2, \ldots, b_p$. The partial derivatives are

$$\frac{\partial S}{\partial a} = \sum \frac{\partial S}{\partial \hat{y}_i} \frac{\partial \hat{y}_i}{\partial a} = \sum -2(y_i - \hat{y}_i)(1) = -2 \sum (y_i - \hat{y}_i) = 0$$

and

$$\frac{\partial S}{\partial b_j} = \sum \frac{\partial S}{\partial \hat{y}_i} \frac{\partial \hat{y}_i}{\partial b_j}$$

$$= \sum -2(y_i - \hat{y}_i)(x_{ji})$$

$$= -2 \sum (y_i - \hat{y}_i) x_{ji} = 0, \, j = 1, 2, \ldots, p.$$

These become

$$\sum \hat{y}_i = \sum y_i$$

and

$$\sum x_{ji} \hat{y}_i = \sum x_{ji} y_i, \, j = 1, 2, \ldots, p.$$

Writing $\hat{y}_i = a + b_1 x_{1i} + b_2 x_{2i} + \cdots + b_p x_{pi}$ and simplifying leads to the $p + 1$ simultaneous linear equations. The solutions $a^*, b_1^*, b_2^*, \ldots, b_p^*$ are the least squares estimates, the values that minimize the least squares criterion. This can be written succinctly as

$$(a^*, b_1^*, b_2^*, \ldots, b_p^*) = \arg \min_{a, b_1, b_2, \ldots, b_p} S(a, b_1, b_2, \ldots, b_p).$$

5.1.3 Types of Explanatory Variables

Note that some of the *explanatory variables* could be functions of others; for example, X_2 might be the square of X_1. One variable can be the product of two others, for example, X_3 might be $X_1 X_2$. Further, the explanatory variables need not be strictly numerical; some could be dummy (0,1) variables. In this book the notation $\mathcal{E}[Y \mid x_1, x_2, \ldots, x_p]$, is used even when the values of the explanatory variables are fixed, as in a planned experiment, that is, a *designed experiment.*

An example with numerical explanatory variables is taken up next. Later in the chapter, examples in which one or more of the explanatory variables are binary will be discussed.

5.2 Market Models

The preceding chapter treated the CAPM and the financial analyst's "beta." The CAPM is an example of a *market model.* Market models describe the ROR of an asset in terms of the behavior of the market as a whole. The CAPM attempts to explain this behavior by a single variable, the excess ROR of the market. Generally, the behavior of an asset is reflected not only in the ROR of a market index, but also in other explanatory variables reflecting market characteristics.

Some tests of the CAPM showed that it did not hold in some situations and was often inaccurate or unsuitable in predicting asset values. Roll (1977) stated that the CAPM holds in theory but is difficult to test empirically because stock indexes and other measures of the market are poor proxies for the CAPM variables. This came to be known as "Roll's critique." The CAPM is still widely taught because of its insights into capital markets and because it is sufficient for many important applications. But next we discuss extensions of it.

5.2.1 Fama/French Three-Factor Model

Economists and financial analysts sometimes use the term "factor" for what statisticians call a "variable." (Statisticians use the term "factor" for the explanatory variables in a planned experiment, especially when they are set at discrete levels. They use the term also for the linear combinations of variables estimating latent variables in *factor analysis,* a part of multivariate statistical analysis.)

Fama and French (1992) introduced two other variables ("factors") along with the overall market factor, resulting in a three-factor model. Their work is geared toward assessing the performance of various portfolios, that is, sets of stocks. In that context, R_{it} is the ROR of portfolio i in time period t. The two additional factors, in addition to the market ROR, are called SMB and HML:

SMB, for "small [cap] minus big" (smallness)

HML, for "high [book/price] minus low" (value)

These variables ("factors") SMB and HML measure the excess returns of small caps (stocks of firms with small capitalization, that is, firms that are small in this sense) and "value" stocks over the market as a whole. The *Fama/French model* is of the form

$$\mathcal{E}[R - R_f] \;=\; \alpha + \; \beta_m \,(R_m - R_f) + \beta_s \cdot SMB + \beta_v \cdot HML.$$

Here R is the portfolio's rate of return, R_f is the risk-free return rate, and R_m is the return of the whole stock market. Because of the way they are defined, the corresponding coefficients β_s and β_v take values on a scale of roughly 0 to 1: $\beta_s = 1$ would be a small cap portfolio, $\beta_s = 0$ would be large cap, $\beta_v = 1$ would be a portfolio with a high book/price ratio, etc.

This model is in the form of a multiple regression with three explanatory variables,

$$\mathcal{E}[Y \mid x_1, x_2, x_3] \;=\; \alpha + \beta_1 x_1 + \beta_2 x_2 + \beta_3 x_3,$$

with $Y = R - R_f$, $x_1 = R_m - R_f$, $x_2 = SMB$, and $x_3 = HML$.

5.2.2   Four-Factor Model

Carhart (1997) extended the Fama/French model to include a *momentum factor,* denoted by MOM or UMD, that is, up-minus-down, a ROR for prior-month winners minus a ROR for prior-month losers.

This is a multiple regression model with four explanatory variables,

$$\mathcal{E}[Y \mid x_1, x_2, x_3, x_4] \;=\; \alpha + \beta_1 x_1 + \beta_2 x_2 + \beta_3 x_3 + \beta_4 x_4,$$

with $Y = r - r_f$, $x_1 = r_m - r_f$, $x_2 = SMB$, $x_3 = HML$, and $x_4 = UMD$.

5.3   Models with Numerical and Dummy Explanatory Variables

In this section, models with both numerical and dummy explanatory variables will be discussed.

5.3.1   Two-Group Models

Consider further the example from the preceding chapter on runs of a car with premium or regular gasoline. Introduce a dummy variable g_t, which is 0 for runs with regular and 1 for runs with premium. This is $g_t = 0$ for runs $t = 1, 2, \ldots, 10$ and $g_t = 1$ for $t = 11, 12, 13, 14$. Letting $Y = $ miles and $X = $ gallons, the regression function can be written as

$$\mathcal{E}[Y_t \mid x_t, g_t] = \beta x_t + \Delta\beta\, x_t\, g_t.$$

Letting $\beta_{reg} = \beta$ and $\beta_{prem} = \beta_{reg} + \Delta\beta$, this is

$$
\begin{aligned}
\mathcal{E}[Y_t \mid x_t, 0] &= \beta x_t = \beta_{reg}\, x_t, \ \text{for } g_t = 0 \\
\mathcal{E}[Y_t \mid x_t, 1] &= \beta x_t + \Delta\beta\, x_t\,(1) = \beta x_t + \Delta\beta\, x_t \\
&= (\beta + \Delta\beta)\, x_t = \beta_{prem} x_t, \ \text{for } g_t = 1.
\end{aligned}
$$

To fit the model, in a spreadsheet enter columns for y, x, g, and $x \times g$. Then in software fit the regression of y on x and $x \times g$, using the no-constant ("constant is zero") option. The output will include the estimates of β and $\Delta\beta$ and the t values and p values. The hypothesis $\mathrm{H_0}: \ \beta_{prem} = \beta_{reg}$ is equivalent to $\mathrm{H_0}: \ \Delta\beta = 0$.

5.3.2   Other Market Models

5.3.2.1   Two Betas

Suppose you wanted to take a long position—buy and hold—on a stock. What sort of stock would you like? One that goes up, you say—one that goes up whether the market goes up or down, that is, whether the market is in a Bull state or a Bear state. This would mean that stock would need a positive beta in a Bull market and a negative beta in a Bear market. Now, you probably will not find a stock that behaves that way, but you might find some stocks that have a large beta in a Bull market and a small beta in a Bear market. Anyway, right away, this leads to a concept of *two* betas. A dummy variable can be included to label each month as Bull or Bear. Then a multiple regression model can be fit, including this variable. The simplest way of defining such a variable is Bull (1) if the market ROR was positive and Bear (0) if the market ROR was negative. This is called the "Up-Down" (U/D) method. method. It is not exactly what most analysts mean by Bull and Bear states, because with the U/D method the state can shift in a single month; other ways of defining Bull and Bear states will be mentioned later in Chapter 9 on regime switching.

In studying rates of return of forty-nine mutual funds for a six-year period 1966 through 1971, Alexander and Stover (1980) included a dummy variable in the model to determine whether the beta coefficient for an individual mutual fund depends on whether the market is moving generally upward (Bull market) or generally downward (Bear market). A six-year period was used, but the conventional period for estimating "beta" is five years of monthly data. (See also McClave and Benson (1994), pp. 681–682 and 686.)

Following this idea, some data (hypothetical but consistent with recent real data) are shown in Table 5.1. The RORs shown are excess RORs; the risk-free rate (three-month Treasury bill rate TB3MS) has been subtracted. (By the way, although over the past several decades the risk-free rate has varied from above 15% down to about 0.01%, in recent years it has been very low.) The asset is a mutual fund. The Up-Down (U/D) Bull/Bear indicator is shown in Table 5.1. Several models will be estimated: the CAPM without constant, the CAPM with constant, and models with the U/D Bull versus Bear indicator.

TABLE 5.1

Excess RORs, with Bull/Bear Indicator

	Month	Market	Fund	Bull		Month	Market	Fund	Bull
5	J	2.16%	2.45%	1	3	J	−1.13%	−1.80%	0
yrs	F	−0.32%	-0.50%	1	yrs	A	1.07%	1.52%	1
ago	M	0.73%	0.75%	1	ago,	S	−9.61%	-9.98%	0
	A	0.82%	1.35%	1	cont'd	O	−18.62%	-18.24%	0
	M	−3.53%	−3.75%	0		N	−7.80%	−8.80%	0
	J	−0.39%	−0.78%	1		D	0.78%	1.35%	1
	J	0.09%	0.22%	1	2	J	−8.97%	−9.69%	0
	A	1.69%	1.96%	1	yrs	F	−11.67%	−9.11%	0
	S	2.03%	2.16%	1	ago	M	8.18%	7.58%	1
	O	2.69%	3.13%	1		A	8.96%	8.61%	1
	N	1.22%	1.31%	1		M	5.16%	5.51%	1
	D	0.85%	1.40%	1		J	0.00%	0.31%	1
4	J	0.98%	1.14%	1		J	7.14%	8.69%	1
yrs	F	−2.63%	−2.99%	0		A	3.29%	3.64%	1
ago	M	0.58%	0.71%	1		S	3.50%	4.34%	1
	A	3.83%	4.51%	1		O	−2.00%	−2.37%	0
	M	2.81%	3.27%	1		N	5.57%	4.67%	1
	J	−2.18%	−1.73%	0		D	1.76%	2.91%	1
	J	−3.65%	−4.22%	0	1	J	−3.77%	−4.13%	0
	A	0.93%	0.94%	1	yrs	F	2.80%	4.51%	1
	S	3.19%	4.16%	1	ago	M	5.70%	6.28%	1
	O	1.15%	1.29%	1		A	1.45%	1.32%	1
	N	−4.78%	−3.23%	0		M	−8.57%	−8.55%	0
	D	−1.12%	−0.56%	0		J	−5.55%	−5.81%	0
3	J	−6.54%	−7.32%	0		J	6.64%	7.09%	1
yrs	F	−3.71%	−1.97%	0		A	−4.87%	−5.05%	0
ago	M	−0.70%	−1.36%	0		S	8.38%	8.96%	1
	A	4.54%	4.38%	1		O	3.61%	2.45%	1
	M	0.92%	1.92%	1		N	−0.24%	0.59%	0
	J	−9.14%	−9.15%	0		D	6.31%	7.31%	1

The first model considered is the CAPM with a constant. Output from the regression of Excess ROR of the mutual fund on that of the market index is shown next. The variable "fund" is the excess ROR of a mutual fund; "market" is the excess ROR of a market index. The value 1.01 of the regression coefficient is the financial analyst's beta. The standard error of fit (square root of the Mean Square for Residual Error, 0.573) is $s = 0.757078$. The output is shown after Table 5.1.

```
Regression Analysis:  Fund versus Market (without constant)

The regression equation is:  fitted value of fund = 1.01 market

Predictor      Coef  SE Coef     t      p
No constant
S&P500      1.00961  0.01870  54.00  0.000
s = 0.757078

Analysis of Variance
Source          DF      SS      MS        F      p
Regression       1  1671.4  1671.4  2916.11  0.000
Residual Error  59    33.8   0.573
Total           60  1705.2
```

Next, the CAPM with a constant (α, representing "differential" ROR) is fit. The output is shown below. Check to see if the beta changes much, and whether the estimate of α is s.d.f.z. (significantly different from zero).

```
Regression Analysis: fund versus market

The regression equation is
fitted value of fund = 0.228 + 1.01 market

Predictor      Coef    SE Coef     t      p
Constant    0.22816    0.09397   2.43  0.018
market      1.01099    0.01797  56.25  0.000

s = 0.727497   R-Sq = 98.2%   R-Sq(adj) = 98.2%

Analysis of Variance
Source          DF      SS      MS        F      p
Regression       1  1674.3  1674.3  3163.57  0.000
Residual Error  58    30.7     0.5
Total           59  1705.0
```

The estimate of beta remained about 1.01, an estimate of 1.01099 instead of the previous 1.00961. (This is too many decimals, anyway: the standard error of estimate of β is about 0.02). The estimate of the constant is s.d.f.z. with a t of 2.43 ($p = .018$). The standard error of fit is about 0.727, an improvement compared to about 0.757 for the fit with no constant.

Next, two betas—a Bull beta and a Bear beta—will be estimated, using the U/D variable. First this is done without constants in the model and then with constants.

```
Regression Analysis: fund versus market, market*U/D (without constants)

The regression equation is
fitted value of fund  =  0.981 market  +  0.0820 market*U/D


Predictor      Coef  SE Coef      t      p
No constant
market      0.98092  0.02250  43.59  0.000
market*U/D  0.08202  0.03804   2.16  0.035

s = 0.734705

Analysis of Variance
Source          DF       SS      MS       F      p
Regression       2  1673.93  836.96  1550.53  0.000
Residual Error  58    31.31    0.54
Total           60  1705.24
```

The fitted model is $\hat{fund} = 0.981$ market $+ 0.0820$ market*U/D. This is

$$\hat{fund} = 0.981 \text{ market, for } U/D = 0(\text{Bear market})$$
$$= 1.083 \text{ market, for } U/D = 1(\text{Bull market}).$$

It is interesting to compare the beta with 1; here, it is less than 1 in a Bear market and greater than 1 in a Bull market. This model gives a standard error of fit equal to about 0.735, not quite as low as that (0.727) of the model with an alpha and one beta.

Next, a model with two betas but a single alpha is estimated, and then one with two betas and two alphas.

```
Regression Analysis: fund versus market, market*U/D

The regression equation is
fitted value of fund = 0.173 + 1.00 market + 0.0291 market*U/D


Predictor      Coef  SE Coef      t      p
Constant     0.1727   0.1467   1.18  0.244
market      1.00047  0.02791  35.84  0.000
market*U/D  0.02911  0.05881   0.49  0.623

s = 0.732278   R-Sq = 98.2%   R-Sq(adj) = 98.1%

Analysis of Variance
Source          DF       SS      MS       F      p
Regression       2  1674.45  837.23  1561.32  0.000
Residual Error  57    30.57    0.54
Total           59  1705.02
```

The fitted model is $\hat{\text{fund}} = 0.173 + 1.00$ market $+ 0.0291$ market*U/D. This is

$$\hat{\text{fund}} = 0.173 + 1.00 \text{ market, if U/D} = 0$$
$$= 0.173 + 1.0291 \text{ market, if U/D} = 1.$$

Next, look at results for fitting two betas and two alphas.

```
Regression Analysis: fund versus U-D, market, market*U/D

The regression equation is
fitted value of fund = - 0.090 + 0.394 U-D + 0.971 market + 0.0334 market*U/D

Predictor      Coef  SE Coef      t      p
Constant    -0.0896   0.2525  -0.35  0.724
U-D          0.3936   0.3094   1.27  0.209
market       0.97077  0.03628  26.76  0.000
market*U/D   0.03343  0.05860   0.57  0.571

s = 0.728337   R-Sq = 98.3%   R-Sq(adj) = 98.2%

Analysis of Variance
Source          DF       SS      MS       F       p
Regression       3  1675.31  558.44  1052.71  0.000
Residual Error  56    29.71    0.53
Total           59  1705.02
```

The fitted regression equation is

$$\hat{\text{fund}} = -0.090 + 0.394U/D + 0.971 \text{ market} + +0.0334 \text{ market*U/D}$$

with a standard error of fit of 0.728, not quite as low as the 0.727 obtained with one beta and one alpha. The estimate of the constant is not s.d.f.z.

The differences in goodness-of-fit of these several models are not dramatic, the standard errors of fit ranging from about 0.76 to about 0.72. However, the exercise of fitting shows how to formulate, fit, and compare alternative, competing models, in the case models with one or two betas and one or two alphas. Remember that the perhaps overly simple U/D has been used for the Bull/Bear states. Other possibilities are considered later.

5.3.2.2 More Advanced Models

Two variances. The error variances may differ between Bull and Bear markets. Separate models for Bull and Bear can be fit, and the estimated error variances compared. Larger variance seems to be associated with Bull periods. Another way of allowing different variances is to use one distribution for Bull periods and another for Bear periods, with possibly different variances as well as different means. *Regime switching models* provide a way of doing this.

Regime Switching Models. Studies employing conventional definitions of Bull and Bear are interesting. However, other models allow the data to speak for themselves to determine switches between states. The estimates of the model parameters may or may not then correspond to conventional definitions of Bull and Bear. A *hidden Markov model* (HMM) consists of state-conditional probability functions, for example Normal distributions with different means and different variances, and a matrix of transition probabilities, entry (j, k) of which gives the probability of transition to state k, given that the process was in state j in the preceding time period. Such models are discussed to some extent in Chapter 9.

5.4 Model Building

Suppose there are several competing alternative models. Generally these will be models involving different explanatory variables. These models are to be ranked in some way, or at least it is to be decided which are reasonably good and which are not.

5.4.1 Principle of Parsimony

It takes only so many parameters to fit a dataset. The *principle of parsimony* is that no more parameters than necessary should be used. That is, there must be a balance between model fit and model complexity. A "model-selection criterion" is an aid to achieve this balance.

5.4.2 Model-Selection Criteria

Model-selection criteria provide figures-of-merit for alternative models, that is, they assign scores to the alternative models.

5.4.2.1 Residual Mean Square

The decomposition of sum of squares for any given model is SST = SSR + SSE. The residual mean square is MSE = SSE/DFE = $\text{SSE}/(n-k-1)$, where k is the number of Xs. The standard error of fit used above is just the square root of MSE.

MSE involves both the sample size n and the number k of explanatory variables used. Its square root s, the standard error of the fit, is a kind of average error of prediction across the n cases in the sample. It is the root-mean-square of these errors (except that a divisor of $n-k-1$ rather than n is used in computing this average). MSE is of course a smaller-is-better criterion. MSE, because it involves the number of explanatory variables k, can be useful for comparing models with different numbers of explanatory variables. (SSE, on the other hand, cannot increase when an additional variable is included in the regression, so it is not a suitable model-selection criterion for models involving different numbers of variables, as it would choose the model with all variables included.)

5.4.2.2   Adjusted R-Square

The lack of fit of the model, in terms of sums-of-squares, is $1-R^2 = $ SSE/SST. A degrees-of-freedom adjustment of this consists of replacing SSs with MSs. This defines *adjusted R-square*, R^2_{adj}, through $1 - R^2_{adj} = $ MSE/MST. Ranking models via adjusted R-square is equivalent to ranking them by MSE, because MST is constant across models, so adjusted R-square varies with the model via MSE.

5.4.3   Testing a Reduced Model against a Full Model

Suppose we want to compare a full model with $p+q$ Xs with a *reduced model* containing only the first p Xs. The full model is of the form

$$Y_i = \beta_0 + \beta_1 x_{1i} + \beta_2 x_{2i} + \cdots + \beta_p x_{pi} + \beta_{p+1} x_{p+1,i} + \cdots + \beta_{p+q} x_{p+q,i} + \varepsilon_i;$$

the reduced model, of the form

$$Y_i = \beta_0 + \beta_1 x_{1i} + \beta_2 x_{2i} + \cdots + \beta_p x_{pi} + \varepsilon_i.$$

MSE is one reasonable criterion for making the comparison between models. We have $\text{MSE}_{red} = \text{SSE}_{red}/\text{DFE}_{red}$ and $\text{MSE}_{full} = \text{SSE}_{full}/\text{DFE}_{full}$. It is interesting that $\text{MSE}_{red} < \text{MSE}_{full}$ is equivalent to $F < 1$, where this is the F for testing the reduced model against the full model,

$$F = \frac{(\text{SSE}_{red} - \text{SSE}_{full})/q}{\text{MSE}_{full}}.$$

The reduced model is better according to MSE if $\text{MSE}_{red} < \text{MSE}_{full}$. This is equivalent to $\text{SSE}_{red}/\text{DFE}_{red} < \text{SSE}_{full}/\text{DFE}_{full}$. This can be shown to be equivalent to $F < 1$.

5.4.4   Comparing Several Models

More generally, several models are to be compared. Index these models by $k = 1, 2, \ldots, K$, where K is the number of alternative models (number of models being compared). One way of ranking the models is according to MSE_k.

Note that $\text{MSE}_k = \text{SSE}_k / (n - p_k - 1)$, where p_k is the number of explanatory variables in Model k. In Gaussian linear additive models, SSE is particularly relevant because it relates directly to the maximum likelihood of the model. SSE is, in turn, an ingredient in BIC (Schwarz 1978), an interesting and frequently used model-selection criterion. BIC is a smaller-is-better criterion that is a penalized lack-of-fit criterion. Given K alternative models, indexed by $k = 1, 2, \ldots, K$,

$$\text{BIC}_k = \text{Deviance}_k + \text{Penalty Term}_k = \text{Deviance}_k + (\ln n)m_k.$$

Here, the Penalty Term$_k$ is like a cost of fitting the parameters, m_k being the number of free parameters estimated in Model k, and the Deviance$_k$ is the lack-of-fit -2LL_k, where LL_k is the log of the maximized likelihood of Model k. (See also Kashyap (1981) for a derivation with a few more details.) Note that BIC incorporates the lack-of-fit, the sample size, and the number of parameters.

For Gaussian linear models, the deviance is a function of the residual sum of squares,

$$-2\,\text{LL}_k = n \ln 2\pi + n \ln \text{SSE}_k - n \ln n + n.$$

Ignoring constants that do not vary across models k leaves $n \ln \text{SSE}_k$. Thus, to rank Gaussian models it suffices to compare $n \ln \text{SSE}_k + (\ln n)m_k$ across models, the lowest value indicating the best model.

BIC stands for Bayesian Information Criterion. BIC is derived by expanding $-2 \ln p(k \mid \text{data})$ and integrating out a prior over the distributional parameters. Here, $p(k \mid \text{data})$ is the posterior probability of Model k, $k = 1, 2, \ldots, K$. To put BIC on a probability scale, note then that $-2 \ln p(k \mid \text{data}) \approx \text{Const.}\, p_k \text{BIC}_k$, where p_k is the prior probability of Model k. So, $\ln p(k \mid \text{data}) \approx -\text{Const.}.\, p_k \text{BIC}_k / 2$, and $p(k \mid \text{data}) \approx \text{Const.} \exp[-p_k \text{BIC}_k / 2]$. The constant is Const. $= \sum_{k=1}^{K} p_k \exp(-\text{BIC}_k/2)$, making the probabilities sum to one. Given equal prior probabilities for the K models, the posterior probability of Model k is approximately Const. $\exp(-\text{BIC}_k/2)$, where Const. $= \sum_{k=1}^{K} \exp(-\text{BIC}_k/2)$.

5.4.5 Combining Results from Several Models

It is not necessary to use only the best model. Results from alternative models can be weighted using their posterior probabilities. Let $\hat{y}_i^{(k)}$ be the prediction of Y_i based on Model k. An overall prediction is

$$\hat{y}_i = \sum_{k=1}^{K} p(k \mid \text{data})\, \hat{y}_i^{(k)},$$

where $p(k \mid \text{data})$ is the posterior probability $\Pr(\text{Model } k \mid \text{data})$, approximated via BIC.

5.5 Chapter Summary

Regression analysis attempts to explain the variation in a response variable Y in terms of a set x of explanatory variables.

A *regression function* is the conditional expected value of Y, given x, where $x' = (x_1 x_2 \ldots x_p)$. The conditional expected value is denoted by $\mathcal{E}[Y \mid x]$.

A *linear* regression function takes the form $\alpha + \beta' x$, where $\beta' = (\beta_1\ \beta_2\ \ldots\ \beta_p)$.

A *simple* linear regression function has the form $\alpha + \beta x$. A *multiple* linear regression function has the form $\alpha + \beta_1 x_1 + \beta_2 x_2 + \beta_p x_p$.

An *additive model* takes the form $Y = \mathcal{E}[Y \mid x] + \varepsilon$, where the error ε is *added* to the regression function.

A *statistical* model is stated in terms of variables such as Y, X, ε, for example as $Y = \alpha + \beta X + \varepsilon$. The corresponding *observational* model is stated in terms of values of the variables for n cases, as $Y_i = \alpha + \beta x_i + \varepsilon_i$, $i = 1, 2, \ldots, n$.

The *partial correlation* between X and Y taking account of Z is the correlation between the residuals of X and Y after regression on Z.

Market models attempt to explain the behavior of asset prices in terms of the movement of the market as a whole.

Regimes are states of the market such as Bull and Bear states. The Bull state is associatied with positive rates of return, the Bear state, with negative rates of return. Often, the volatility (in the sense of standard deviation of RORs) is higher in the Bear state.

5.6 Chapter Exercises

5.6.1 Exercises for Two Explanatory Variables

5.1 Dr. Smith wants to decide whether or not to use a standard chemistry test along with her own test to advise students whether to take AP chemistry. For each of $n = 53$ students she has scores on $y = $ final numerical grade in chemistry (mean $= 70$), $x_1 = $ her pretest (mean $= 60$), and $x_2 = $ standard chemistry aptitude test (mean $= 50$). The sums of squares and cross-products of deviations are as follows:

$$
\begin{array}{lll}
a_{11} = 10 & a_{12} = 5 & a_{1y} = 15 \\
 & a_{22} = 10 & a_{2y} = 13 \\
 & & a_{yy} = 38
\end{array}
$$

a. Find the regression equation of Y on X_1 and X_2.

b. Compute SSE, DFE, MSE, and the standard error of estimate.

c. Find the regression equation of Y on X_1 alone.

d. For this regression, compute the residual sum of squares, residual mean square, and standard error of estimate.

e. Compute the standard deviation of Y. Compare it with the two standard errors of estimate found.

5.2 Compute the three pairwise correlations for the three variables of the preceding exercise.

5.3 (continuation) Compute the *partial correlation* between Y and X_2, adjusting for X_1, by means of the formula

$$r_{yx_2.x_1} = \frac{r_{y2} - r_{12}r_{2y}}{(1 - r_{12}^2)^{1/2}(1 - r_{2y}^2)^{1/2}}.$$

Compare this partial correlation with the total correlation between Y and X_2.

5.4 Compute $r_{yx_1.x_2}$ when

a. $r_{y1} = .8, r_{12} = .9, r_{y2} = .8$

b. $r_{y1} = .8, r_{12} = .9, r_{y2} = .9$

c. Briefly interpret the values of the partial correlation in the two cases.

5.5 Suppose Y = housing starts, X_1 = savings-and-loan association mortgage volume, and X_2 = commercial bank mortgage volume. Interpret the values of $r_{yx_1.x_2}$ in the two cases.

5.6 Your car can use either regular or premium gasoline (though possibly not equally efficiently or cleanly). Regular costs $3.199 per gallon; premium, $3.599. You drive 12,000 miles per year. Maintenance costs $300 each time. You will need two maintenances a year with regular, only one with premium. Your car gets 25 mpg with regular, 30 with premium. The total cost is gasoline cost plus maintenance cost. Compare the annual total costs with premium and regular. (This is not a statistical problem but rather asks that estimates be incorporated into a decision problem.)

5.7 Smith's car can use either regular or premium gasoline (though possibly not equally efficiently or cleanly). Regular gasoline costs him $3.099 per gallon; premium, $3.599. He drives 24,000 miles a year. Maintenance costs $250 each time. He will need two maintenances a year with regular, only one with premium. His car gets 20 mpg with regular, 25 with premium. The total cost is gasoline cost plus maintenance cost. Compare Smith's annual total costs with premium and regular.

5.6.2　Mathematical Exercises: Two Explanatory Variables

5.8 Derive estimating equations for the model $y_i = \beta_1 x_{1i} + \beta_2 x_{2i} + \varepsilon_i, i = 1, 2, \ldots, n$, by the heuristic method of reduction of observations shown in the text.

5.9 (continuation) Derive estimating equations for the simple linear regression model $y_i = \beta_0 + \beta x_i + \varepsilon_i$ by setting $\beta_1 = \beta_0, \beta_2 = \beta, x_{1i} = 1$, and $x_{2i} = x_i$ in the preceding exercise.

5.10 (continuation) Construct the F test of H: $\beta_0 = 0$ and $\beta = 1$, that is, test the hypothesis that the true model is $y_i = x_i + \varepsilon_i$.

5.11 (continuation) Construct the F (or t) test of the hypotheis H: $\beta = 1$.

5.12 Derive estimating equations for the model $y_i = \beta_0 + \beta_1 x_{1i} + \beta_2 x_{2i} + \varepsilon_i, i = 1, 2, \ldots, n$, by the heuristic method of reduction of observations.

5.13 The sample *partial correlation* between Y and X_2, adjusting for X_1, denoted by $r_{yx_1.x_2}$ (see also Exercise 4.3), is the simple correlation between the parts of Y and X_2 not related to X_1, that is, between $\tilde{Y}$ and $\tilde{X}_2$, where $\tilde{y} = y - \hat{y}, \tilde{x}_2 - \hat{X}_2$, and $\hat{y}|x_1 = \bar{y} + b_y 1 x_1', \hat{x}_2|x_1 = \bar{x}_2 + b_{21} x_1'$. Show that

$$r_{yx_2 \cdot x_1} = \frac{r_{y2} - r_{y1} r_{12}}{(1 - r_{y1}^2)^{1/2}(1 - r_{12}^2)^{1/2}}.$$

5.14 Show that

$$1 - R_{y \cdot x_1 x_2} = (1 - r_{y1}^2)(1 - r_{yx_2.x_1}).$$

5.15 The *stage-wise* estimator formed by regressing Y on X_1, and then the residuals on X_2, is

$$\hat{\beta}_2 = \frac{\sum (x_2 - \bar{x}_2)(y - y|x_1)}{\sum (x_2 - \bar{x}_2)^2},$$

where

$$y|x_1 = \bar{y} + b_{y1}(x_1 - \bar{x}_1).$$

The stagewise procedure is useful in exploratory data analysis when one is studying the relationship between Y and X_1 and then realizes that the residuals may be correlated with another variable X_2.

　　a. Assuming the model $Y = \beta_0 + \beta_1 x_1 + \beta_2 x_2 + \varepsilon$, compute the mean of the estimator f $\tilde{\beta}_2$.

　　b. When is this estimator unbiased?

5.16 (continuation) Compute the mean squared error of this estimator.

5.17 (continuation) The mean squared error of b_2, the LSE, is

$$\sigma^2 / \sum [x_2 - (x_2 \,|\, x_1)]^2.$$

When is the mean squared error of the stagewise estimator less than that of the LSE, that is, for what parameter values? (See Goldberger, 1961.)

5.6.3 Mathematical Exercises: Three Explanatory Variables

5.18 What is SSE_{red} for testing H: $\beta_1 = 2, \beta_2 = 3, \beta_3 = 4, \beta_0 = 5$? What are the numbers of degrees of freedom for F?

5.19 What is SSE_{red} for testing H: $\beta_0 = 0$. What are the numbers of degrees of freedom for F?

5.20 What is SSE_{red} for testing H: $\beta_0 = 0$? What are the numbers of degrees of freedom for F?

5.21 What is SSE_{red} for testing H: $\beta_0 = 0, \beta_1 = 1, \beta_2 = 1, \beta_3 = 1$? What are the numbers of degrees of freedom for F?

5.22 Show that

$$\{(\beta_0, \beta_1, \beta_2, \beta_3) : (\sum [y - (\beta_0 + \beta_1 + \beta_2 + \beta_3)]^2 / s^2 \le F_{4,n-4}(.05)\},$$

where $s^2 = \text{SSE}_{full}/(n-4)$, is a 95% confidence region for $(\beta_0, \beta_1, \beta_2, \beta_3)$.

5.23 What is SSE_{red} for testing H: $\beta_1 = \beta_2 = \beta_3 = 17$? What are the numbers of degrees of freedom for F?

5.24 What is SSE_{red} for testing H: $\beta_1 = \beta_2 = \beta_3$? What are the numbers of degrees of freedom for F? *Hint:* The hypothesis is H: $\beta_1 = \beta_2 = \beta_3 = \beta$, where β is unspecified. That is, under H, there is still one free parameter.

5.6.4 Exercises on Subset Regression

5.25 Consider four variables having the correlation matrix of Table 5.2. In what order will a forward selection procedure, based at each stage on maximum partial correlation (that is, on maximum reduction of residual sum of squares), introduce the Xs into the regression equation?

5.26 Consider four variables having the correlation matrix of Table 5.3. (This matrix differs from that of the preceding exercise only in that r_{12} is .6 here, instead of .8.)

In what order will a forward selection procedure, based at each stage on maximum partial correlation (that is, on maximum reduction of residual sum of squares), introduce the Xs into the regression equation ?

5.27 Outline a procedure for the situation of choosing items for a test in Exercise 5.1.

TABLE 5.2
Correlation Matrix of Four Variables

	x_1	x_2	x_3
y	.6	.4	.3
x_1		.8	.1
x_2			.5

TABLE 5.3
Correlation Matrix of Four Variables

	x_1	x_2	x_3
y	.6	.4	.3
x_1		.6	.1
x_2			.5

5.6.5   Mathematical Exercises: Subset Regression

5.28 How many ways are there to choose 20 items from among 40?

5.29 Prove that introducing an additional X into the regression equation cannot increase SSE.

5.30 When an additional X is introduced into a regression equation, how great must the reduction in SSE be in order that there also be a reduction in MSE?

5.31 Derive the likelihood L for the classical Normal model.

5.32 A special case of model selection occurs when the number K of alternative models is just 2. Then the procedure is hypothesis testing. Use BIC to test $H_0 : \beta_2 = 0$ in the model where $Y = \beta_0 + \beta_1 x_1 + \beta_2 x_2 + \varepsilon$, given the classical Normal assumptions on the errors.

5.33 (continuation) What is the level of the the test? What is the numerical value of the level if $n = 25$?

5.7 Bibliography

Alexander, Gordon J., and Stover, Roger D. (1980). Consistency of mutual fund performance during varying market conditions. *Journal of Economics and Business*, **32**, 219–226.

Black, Fischer (1972). Capital market equilibrium with restricted borrowing. *Journal of Business*, **45**, 444-454.

Carhart, Mark (1997). On persistence in mutual fund performance. *Journal of Finance*, **52(1)**, 57–82.

Fama, Eugene F., and French, Kenneth R. (1992). The cross-section of expected stock returns. *Journal of Financial Economics*, **33**, 3–56.

Goldberger, Arthur S. (1961). Note on stepwise least squares. *Journal of the American Statistical Association*, **56**, 105–110.

Kashyap, Rangasami L. (1982). Optimal choice of AR and MA parts in autoregressive moving average models. *IEEE Transactions on Pattern Analysis and Machine Intelligence*, **4**, 99–104.

McClave, James T., and Benson, P. George (1994). *Statistics for Business and Economics. 6th ed.* Dellen (Macmillan), Englewood Cliffs, NJ.

Roll, Richard (1977). A critique of the asset pricing theory's tests. Part I: On past and potential testability of the theory. *Journal of Financial Economics*, **4(2)**, 129–176.

Schwarz, Gideon (1978). Estimating the dimension of a model. *Annals of Statistics*, **4**, 461–464.

Part III

PORTFOLIO ANALYSIS

6

Mean-Variance Portfolio Analysis

CONTENTS

6.1   Introduction

This chapter considers statistical characteristics of sets of assets, especially their means, variances, and correlations. . A set of assets held by an investor is called the investor's *portfolio*. The subject of this and the next chapter is *portfolio analysis*. This means the attempt to optimize the combination of assets, based on available data. Portfolios of m risky stocks are considered, and a combined portfolio of m risky stocks and a risk-free asset is considered.

A section on matrices and vectors is included in Appendix 6A and Appendix A at the end of the book. These are optional; however, it is recommended that persons interested in statistical finance become acquainted with vectors and matrices and operations on them.

As stated above, a *portfolio* is a set of assets, such as stocks, held by an investor. *Portfolioselection* refers to the choice of stocks for the portfolio. *Portfolio allocation* refers to the choice of weights of those stocks in the portfolio, where the weight of a stock is the proportion of total investment put into that stock. Here the focus will be on allocation, with a view toward choosing weights that give a good portfolio rate of return and low variability. Much of the chapter concerns formulas for the mean and variance of portfolio rate of return (ROR), based on the means and variances of the RORs of the individual stocks, and their correlations.

The reader may wish to review the notation and definitions given earlier in Chapters 2 and 3, the introduction to financial data.

Characteristics of portfolio ROR, R_p, will be studied in terms of the RORs R_i of the stocks in the portfolio. When the time period t is to be emphasized, the notation R_{pt} is used for portfolio ROR at time t and R_{it} for the ROR of asset i at time t. Related notation is given in Table 6.1.

Now let s_t denote the value of the portfolio at time t. (In terms of an

TABLE 6.1
Portfolio Quantities at Time t

Asset, i	Share price, P_{it}	No. of Shares, n_i	Amount, a_i	Weight, w_i
1	P_{1t}	n_1	$a_{1t} = n_1 P_{1t}$	$w_1 = a_{1t}/s_t$
2	P_{2t}	n_2	$a_{2t} = n_2 P_{2t}$	$w_2 = a_{2t}/s_t$
$\vdots$	$\vdots$	$\vdots$	$\vdots$	$\vdots$
m	P_{mt}	n_m	$a_{mt} = n_m P_{mt}$	$w_m = a_{mt}/s_t$

$$s_t = \sum_{i=1}^{m} a_{it}$$

individual investor, this is sometimes called the investor's "wealth.") Letting a_{it} be the amount in stock i at time t, the total value is $s_t = \sum_{i=1}^{m} a_{it}$. The rate of return of the portfolio from time $t-1$ to time t is $R_{pt} = (s_t - s_{t-1})/s_{t-1}$. But this is a linear combination of the rates of return of the individual assets. To see this, write

$$s_t - s_{t-1} = \sum_{i=1}^{m} a_{i,t} - \sum_{i=1}^{m} a_{i,t-1} = \sum_{i=1}^{m} (a_{i,t} - a_{i,t-1}).$$

Because at time $t-1$ the m assets are held in proportions $w_i = a_{i,t-1}/s_{t-1}$, $i = 1, 2, \ldots, m$, the portfolio ROR is

$$
\begin{aligned}
R_{pt} &= (s_t - s_{t-1})/s_{t-1} \\
&= \sum_{i=1}^{m} (a_{it} - a_{i,t-1})/s_{t-1} \\
&= \sum_{i=1}^{m} [(a_{it} - a_{i,t-1})/a_{i,t-1}](a_{i,t-1}/s_{t-1}) \\
&= \sum_{i=1}^{m} [(a_{i,t} - a_{i,t-1})/a_{i,t-1}]w_i \\
&= \sum_{i=1}^{m} w_i R_{it}.
\end{aligned}
$$

Note that, of course, the rate of return in terms of the share prices P_i is the same as the rate of return in terms of the amounts a_i. The ROR of asset i is

$$
\begin{aligned}
(a_{i,t} - a_{i,t-1})/a_{i,t-1} &= (n_i P_{i,t} - n_i P_{i,t-1})/(n_i P_{i,t-1}) \\
&= (P_{i,t} - P_{i,t-1})/P_{i,t-1} \\
&= R_{i,t}.
\end{aligned}
$$

Portfolio ROR is the linear combination of individuals RORs when ordinary ROR is used and is approximately that linear combination if continuous RORs are used. That is, $R_{pt} = \sum_{i=1}^{m} w_i R_{it} \approx \sum_{i=1}^{m} w_i r_{it}$, because $r_{it} \approx R_{it}$.

6.1.1 Mean-Variance Portfolio Analysis

Bi-criterion portfolio analysis, or *mean-variance* analysis, uses the mean and variance of portfolio ROR. A good portfolio is one with a good combination of portfolio mean ROR, μ_p and variance σ_p^2 of portfolio ROR, namely, relatively high mean and relatively low variance. The possible combinations are usually represented in a plot of mean versus standard deviation or mean versus variance.

A parenthetical note on computation. The plot in Figure 6.1 was made by sampling triplets of weights (a, b, c) for three stocks. Weight a was set at .000(.001)1.000, that is, from .000 to 1.000 in steps of .001 (1,001 values). For each value of a, two values of b were obtained. To insure a representative sample, the values of b were sampled in antithetic pairs. In this case, that means that the first was sampled uniformly in the interval $(a, 1)$. Then, paired with that value of b, the value $(1 - a - b)$ was included. Then of course c was $1 - a - b$. For example, if $a = .200$, then b was chosen uniformly in the interval $(.200, 1)$. Say that b was .350. Then c would be $1 - .200 - .350 = .450$. Then, moving on to $a = .201$, the value of b would be taken as $1 - .201 - .350 = .649$. The next pair of b values would be for $a = .202$ and .203.

The *efficient frontier* consists of the lowest possible standard deviation for any given value of the mean, and the highest possible value of the mean for any given value of the standard deviation. So to obtain the efficient frontier graphically, there are two equivalent procedures producing it. One can find the lowest possible standard deviation for any given value of the mean, or the highest possible value of the mean for any given value of the standard deviation

To find the lowest possible standard deviation for any given value of the mean, start at the left (horizontal axis) and move right until you just reach the set of feasible points. The standard deviation at that point is the lowest

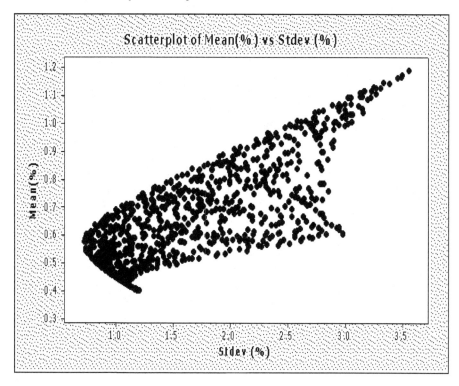

FIGURE 6.1
Mean versus standard deviation

standard deviation for that given value of the mean. This point (σ, μ) is called an efficient point. The set of all such points (upper right boundary of the feasible set) is called the *efficient boundary* or *efficient frontier*.

To find the highest mean for any given value of the standard deviation, start with a given value of σ along the horizontal axis and go as high as possible, to the top of the feasible set, getting the highest possible value of μ at that value of σ.

The good direction is the northwest: high mean (north) and low standard deviation (west). If an additional risky asset is added to the selection, the feasible set will move to the northwest.

6.1.2 Single-Criterion Analysis

It would be useful to work in terms of criteria that combine return and risk, that is, mean and variance. Various such single criteria are discussed. The Sharpe ratio of a portfolio is discussed later in this chapter, as is Value-at-Risk (VaR). A utility-based single criterion involving both mean and variance is discussed in the next chapter.

6.2 Two Stocks

Next the cases of two stocks and three stocks are treated in some detail.

First consider a portfolio of just two stocks, A and B (Table 6.2). Represent their RORs by random variables Xx and Y, with means μ_x, μ_y; and standard deviations σ_x, σ_y. The correlation is ρ_{xy}, the covariance $\sigma_{xy} = \rho_{xy}\sigma_x\sigma_y$. That is, knowing the standard deviations and the correlation, the covariance can be computed. Also, knowing the standard deviations is the same as knowing the variances, as each variance is the square of the corresponding standard deviation. The weights on the two stocks will be denoted by a and b ($a + b = 1$). If \$100,000 is to be invested, and a is .6, then \$60,000 will be put in Stock A, and the other 40 K\$ in Stock B.

TABLE 6.2
Two Stocks. Format of Table of RORs

Month	Stock A	Stock B
1	0.5 %	1.6 %
2	−0.3 %	1.4 %
⋮	⋮	⋮
n	0.2 %	0.9 %

From the statistics in Table 6.3, it appears that Stock A has a moderate ROR, and B has a higher ROR but is riskier. There is a negative correlation that is fairly large in size.

To be good, a portfolio should have high expected return, relative to its risk, and small loss probability, where by loss probability is meant the probability that the portfolio ROR will be negative. We compute this loss probability for Stock A alone, Stock B alone, and a mix of the two stocks.

If the variable X is taken as having a Normal distribution, what is the probability of loss with Stock A alone, that is, what is $\Pr\{X < 0\}$? The standardized value of $x = 0$ is $z = (0 - 0.4)/0.8 = -0.5$, so $\Pr\{x < 0\} = \Pr\{z < -0.5\} \approx .309$.

What is the probability of loss with Stock B alone; that is, assuming Normality, what is $\Pr\{(Y < 0)\}$? Here, $z = (0 - 1.5)/4.5 = -1/3 \approx -0.333$, and the probability to the left of that value under the standard Normal curve is about .370.

TABLE 6.3
Statistics of RORs of Two Stocks

Stock	A	B
ROR	X	Y
Mean ROR	0.4 %	1.5 %
Std.Dev. of ROR	0.8 %	4.5 %
Correlation of RORs	− .6	

Weights a and b are to be chosen to give a portfolio with good characteristics, in particular, a high mean ROR and a low variance of ROR. The portfolio ROR is $R_p = aX + bY$. Formulas for the mean and variance of R_p are needed.

6.2.1 Mean

The formula for the expected value of R_p is

$$\mathcal{E}[R_p] = \mathcal{E}[aX + bY] = a\mathcal{E}[X] + b\mathcal{E}[Y],$$

or, denoting $\mathcal{E}[R_p]$ by the abbreviated notation μ_p, this is $\mu_p = a\mu_x + b\mu_y$.

6.2.2 Variance

The variance $\mathcal{V}[x]$ of a r.v. X is $\mathcal{V}[X] = \mathcal{E}[(X - \mu_x)^2]$, that is, the variance is the expected value of the squared deviation from the mean.

The square root of the variance is the standard deviation, SD[X], or σ_x. The variance is often denoted by σ_x^2.

6.2.3 Covariance and Correlation

The *covariance* of X and Y is $\mathcal{E}[(X - \mu_x)(Y - \mu_y)]$, that is, the covariance is the expected value of the cross-product of deviations from the mean. It is denoted also denoted by $\mathcal{C}[X, Y]$.

The *correlation coefficient,* or, more simply, the correlation, of variables X and Y, denoted by ρ_{xy}, is their covariance, divided by the product of their standard deviations,

$$\text{Corr}(X,Y) = \frac{C[X,Y]}{\text{SD}[X]\text{SD}([Y]},$$

that is,

$$\rho_{xy} = \frac{\sigma_{xy}}{\sigma_x \sigma_y},$$

where σ_x or $\text{SD}[X]$ is the standard deviation of x, and similarly for Y.

The correlation is a *dimensionless* (unitless) quantity. The units in its numerator are cancelled by the units in its denominator. The range of the correlation coefficient is from -1 to +1.

By reversing the above formula, the covariance is expressed as the product of the correlation and the standard deviations by reversing the above formula, that is,

$$C[x,y] = \text{Corr}[x,y]\,\text{SD}[x]\,\text{SD}[y],$$

or

$$\sigma_{xy} = \rho_{xy}\,\sigma_x\,\sigma_y.$$

The sample variance of X is denoted by s_x^2; the sample variance of Y, by s_y^2. The **sample covariance** is denoted by s_{xy}.

The covariance is related to the regression coefficient and to the correlation coefficient; the covariance is the numerator of both. The coefficient of x in the regression of y on x is s_{xy}/s_x^2. The sample correlation coefficient is

$$r_{xy} = \frac{s_{xy}}{s_x s_y}.$$

6.2.4   Portfolio Variance

Expressions for portfolio variance are developed. First, the variance of a sum or a difference are discussed.

6.2.4.1   Variance of a Sum; Variance of a Difference

Fact. $V[U+V] = V[U] + V[V] + 2C[U,V].$

Remark. When the variance of a sum or difference of two variables is computed, the covariance occurs in the cross-product term.

Proof: Let S be the sum, that is, $S = U + V$. Then

$$
\begin{aligned}
V[S] &= \mathcal{E}[(S-\mu_s)^2] \\
&= \mathcal{E}[[(U+V)-\mu_{u+v}]^2] \\
&= \mathcal{E}[[(U-\mu_u)+(V-\mu_v)]^2] \\
&= \mathcal{E}[(U-\mu_u)^2 + (V-\mu_v)^2 + 2(U-\mu_u)(V-\mu_v)] \\
&= \mathcal{E}[(U-\mu_u)^2] + \mathcal{E}[(V-\mu_v)^2] + 2\mathcal{E}[[(U-\mu_u)(V-\mu_v)]] \\
&= V[U] + V[V] + 2C[U,V],
\end{aligned}
$$

where the covariance is $C[U,V] = \mathcal{E}[(U - \mu_u)(V - \mu_v)]$.

Fact. $V[U - V] = V[U] + V[V] - 2C[U,V]$.
Combining the two preceding facts into one statement gives the
Fact. $V[U \pm V] = V[U] + V[V] \pm 2C[U,V]$.

6.2.4.2 Portfolio Variance

The sum and difference are particular linear combinations. Now consider any linear combination V of X and Y, $\quad V = aX + bY$. The expected value of V is, using $\mathcal{E}[aX] = a\mathcal{E}[X]$ and $\mathcal{E}[bY] = b\mathcal{E}[Y]$,

$$\mathcal{E}[V] = \mathcal{E}[aX + bY] = a\,\mathcal{E}[X] + b\,\mathcal{E}[Y] = a\,\mathcal{E}[X] + b\,\mathcal{E}[Y].$$

Fact. $V[aX + bY] = a^2 V[X] + b^2 V[Y] + 2\,ab\,C[X,Y]$.

The standard deviation, SD$[V]$, that is, SD$[aX + bY]$, is of course the square root of the variance $V[aX + bY]$. This standard deviation is denoted also by σ_v or σ_{ax+by}.

Variance as a function of the mean in the case of two stocks. Given values of $\mu_x, \mu_y, \sigma_x, \sigma_y$, and ρ_{xy}, the weight a can be written in terms of μ_p. This expression for a can be substituted into the expression for σ_p^2. Then σ_p^2 can be written in terms of μ_p. The result is of the form $\sigma_p^2 = A\mu_p^2 + B\mu_p + C$, a parabola.

6.2.5 Minimum Variance Portfolio

It is interesting to find the *minimum-variance portfolio*, although this portfolio may not have a good mean ROR.

To minimize the variance, a Lagrange multiplier (see Appendix C) can be used to include the condition $a + b = 1$, or the variance can be written as a function of the weight a. This is a quadratic, that is, an expression of the form $Ax^2 + Bx + C$. When $A > 0$, this function has a minimum at $x^* = -B/2A$. This fact can be used, or the function can be differentiated with respect to a, setting the derivative equal to zero, and solving. The variance is $a^2\sigma_x^2 + b^2\sigma_y^2 + 2ab\sigma_{xy}$, or as a function of a, say

$$V(a) = a^2\sigma_x^2 + (1-a)^2\sigma_y^2 + 2a(1-a)\sigma_{xy}.$$

The derivative is

$$V'(a) = 2\sigma_x^2 - 2\sigma_y^2(1-a) + 2\sigma_x y - 4\sigma_{xy}a.$$

Setting this equal to zero gives

$$\sigma_x^2 - \sigma_y^2(1-a) + \sigma_{xy} - 2\sigma_{xy}a = 0, \text{ or } (\sigma_x^2 + \sigma_y^2 - 2\sigma_{xy})a - \sigma_y^2 + \sigma_{xy} = 0,$$

so the variance-optimal value of a is

$$a^* = (\sigma_y^2 - \sigma_{xy})/(\sigma_x^2 + \sigma_y^2 - 2\sigma_{xy}).$$

Note that this is $C[Y, Y - X]/V[Y - X]$, the coefficient of regression of Y on $Y - X$. The optimal value of b is $b^* = 1 - a^* = (\sigma_x^2 - \sigma_{xy})/(\sigma_x^2 + \sigma_y^2 - 2\sigma_{xy})$. An expression for the minimum-variance weights in the general case of m stocks is derived in Appendix 6A.

6.3   Three Stocks

Next, consider three stocks: A, B, and C (Table 6.4).

TABLE 6.4
Format of Table of RORs for Three Stocks

Stock	A	B	C
ROR	x	y	z
Month			
1	0.5%	1.6 %	−0.1%
2	−0.3%	1.4 %	0.2%
⋮	⋮	⋮	⋮
n	0.2%	0.9 %	1.2%
Mean	0.4%	1.1%	0.9%
Std.Dev.	0.5%	3.1%	4.2%
Correlations			
Stock A		$r_{xy} = +.2$	$r_{xz} = +.3$
Stock B			$r_{yz} = -.7$

Denote the RORs by X, Y, Z. Let the weights be a, b, c, where $a+b+c = 1$. The portfolio ROR is $R_p = aX + bY + cZ$. The expected value is

$$\mathcal{E}[R_p] = a\mu_x + b\mu_y + c\mu_z.$$

The variance is

$$V[R_p] = a^2\sigma_x^2 + b^2\sigma_y^2 + c^2\sigma_z^2 + 2ab\sigma_{xy} + 2ac\sigma_{xz} + 2bc\sigma_{yz}.$$

The covariances are $\sigma_{xy} = \rho_{xy}\sigma_x\sigma_y$, $\sigma_{xz} = \rho_{xz}\sigma_x\sigma_z$, $\sigma_{yz} = \rho_{yz}\sigma_y\sigma_z$. The means, standard deviations, and correlations are estimated by their sample analogs, the sample means, sample standard deviations, and sample correlations. In this example, the sample correlation of x and y is low and positive, that of x and z is low and positive, and that of y and z is negative and large in size.

6.4 m Stocks

Next, consider m stocks, indexed by $i = 1, 2, \ldots, m$, with RORs $R_1, R_2, \ldots, R_m$, or, including the time t in the notation, $R_{1t}, R_{2t}, \ldots, R_{mt}$ at time t (Table 6.5). .

TABLE 6.5
Format of Table of RORs R_{it}, $i = 1, 2, \ldots, m$; $t = 1, 2, \ldots, n$.

Stock	1	2	$\cdots$	m
ROR	R_1	R_2	$\cdots$	R_m
Month				
1	R_{11}	R_{21}	$\cdots$	R_{m1}
2	R_{12}	R_{22}	$\cdots$	R_{m2}
$\vdots$	$\vdots$	$\vdots$	$\cdots$	$\vdots$
n	R_{1n}	R_{2n}	$\cdots$	R_{mn}
Mean	$\bar{R}_1$	$\bar{R}_2$	$\cdots$	$\bar{R}_m$
Std.Dev.	s_1	s_2	$\cdots$	s_m

Let the weights be $a_1, a_2, \ldots, a_m$, where $\sum_{i=1}^{m} a_i = 1$. The portfolio ROR is $R_{R_p} = \sum_{i=1}^{m} a_i R_i$. The expected value is

$$\mathcal{E}[R_p] = \sum_{i=1}^{m} a_i R_i = \sum_{i=1}^{m} a_i \mathcal{E}[R_i] = \sum_{i=1}^{m} a_i \mu_{R-i}.$$

The variance is

$$\mathcal{V}[R_p] = \sum_{i=1}^{m}\sum_{j=1}^{m} a_i a_j \mathcal{C}[R_i, R_j] = \sum_{i=1}^{m} a_i^2 \sigma_{R_i}^2 + 2\sum_{i=1}^{m}\sum_{j=i+1}^{m} a_i a_j \sigma_{R_i R_j}.$$

The covariances are $\sigma_{R_i R_j} = \rho_{R_i R_j} \sigma_{R_i} \sigma_{R_j}$. The means, standard deviations, and correlations are estimated by their sample analogs, the sample means, sample standard deviations, and sample correlations r_{ij}, $i,j = 1, 2, \ldots, m$.

6.5 m Stocks and a Risk-Free Asset

The preceding has concerned portfolios consisting of risky assets. Figure 6.1 shows possible (σ, μ) pairs for such a portfolio. More generally, a portfolio can be a combination of a *risk-free asset,* such as an interest-bearing savings account or certificate of deposit, and a risky portfolio. The ROR of the risk-free asset will be denoted by R_f. The risk-free rate can vary from time to time, so we write R_{ft}. It is usually considered as known in advance (specified); it is a fixed, not a random, variable in the calculations. That is, at the end of period $t - 1$, we know the value of R_{ft}.

Let w_f denote the weight on the risk-free asset and w_r the weight on the risky portion ($w_f + w_r = 1$), and let R_r denote the ROR of the risky part (the subscript r in w_r and R_r denoting "risky"). The ROR of the whole combination portfolio, including both the risk-free and the risky portions, is

$$R_p = w_f R_f + w_r R_r = (1 - w_r) R_f + w_r R_r = R_f + w_r (R_r - R_f).$$

Let μ_r denote the mean ROR of the risky portion. Then the expected value of portfolio ROR is

$$
\begin{aligned}
\mathcal{E}[R_p] &= \mathcal{E}[w_f R_f + w_r R_r] \\
&= w_f R_f + w_r \mu_r \\
&= (1 - w_r) R_f + w_r \mu_r \\
&= R_f + w_r (\mu_r - R_f).
\end{aligned}
$$

The variance of portfolio ROR is, remembering that R_f is taken as fixed, not random,

$$\mathcal{V}[R_p] = \mathcal{V}[w_f R_f + w_r R_r] = \mathcal{V}[w_r R_r] = w_r^2 \sigma_r^2.$$

The standard deviation $\text{SD}[R_p]$ is $\sigma_p = w_r \sigma_r$.

6.5.1   Admissible Points

A point (s, m) in the (σ, μ)-plane is *inadmissible* if there is another point (s', m') such that $s' \leq s$ and $m' \geq m$, at least one of these inequalities being strict. In this case, the point (s', m') is said to *dominate* the point (s, m). The point (s, m) is *admissible* if there is no point that dominates it.

6.5.2   Capital Allocation Lines

Now consider a risky portfolio with standard deviation s and mean m, so that it is represented by the point (s, m) in the (σ, μ)-plane. Consider a portfolio that is a combination of this risky portfolio and a risk-free asset, in proportions w_r and w_f. As w_r varies, a line is traced out in the (σ, μ)-plane. This line is called the *capital allocation line* (CAL) corresponding to the given point (s, m). Any given risky portfolio generates such a line. The intersection of any of these lines with the feasible set $\mathcal{F}$ consists of inadmissible points, because, given a point on any such line, there is a point above it, that is, a point corresponding to a portfolio that has the same standard deviation but a higher mean. The one exception to this is the line tangent to the northwest part of $\mathcal{F}$. This line is called the *tangency line*.

It can be shown that the choice of values of w_f and w_r is separate from the problem of allocation to assets in the risky portion. It depends upon the investor's level of risk aversion. One way of taking all of this into account at the same time is considered in Chapter 7.

6.6   Value-at-Risk

The Value-at-Risk (VaR) at level .05 is the value below which there is only a probability of .05. This is the fifth percentile of the probability distribution of ROR. The probability is .95 of having a value greater than the VaR. Here we study this concept in terms of ROR, R_p. Then the .05 VaR is defined by $\Pr\{R_p \leq \text{VaR}\} = .05$, that is, $\Pr\{R_p > \text{VaR}\} = .95$. More generally, the level α VaR, say VaR_α, is defined by $\Pr\{R_p \leq \text{VaR}_\alpha\} = \alpha$. It is the 100α-th percentile of the distribution of R_p.

6.6.1   VaR for Normal Distributions

Under an assumption that R_p has a Normal distribution, the VaR, like any percentile, depends only upon the mean μ and standard deviation σ. Then

$$
\begin{aligned}
\Pr\{R_p > \text{VaR}\} &= \Pr\{(R_p - \mu)/\sigma > (\text{VaR} - \mu)/\sigma\} \\
&= \Pr\{Z > (\text{VaR} - \mu)/\sigma\} = .95,
\end{aligned}
$$

where Zz has the standard Normal distribution. But $\Pr\{Z > -1.645\} = .05$. Therefore, we set $(\text{VaR} - \mu)/\sigma$ equal to -1.645. This gives $\text{VaR} = \mu - 1.645\sigma$ as the 95% value of VaR. For example, if μ is 0.5% per month and σ is 1% per month, this is $\text{VaR} = 0.5 - 1.645(1) = -1.145\%$. It is unlikely (5% chance) that the ROR will be less than -1.145 % per month.

The set of portfolios having .05-level VaR less than a given value VaR_0 is the set having $\mu - 1.645\sigma \geq \text{VaR}_0$. In the (σ, μ)-plane, these are the portfolios represented by the points above the line $\mu = \text{VaR}_0 + 1.645\sigma$.

6.6.2 Conditional VaR

It is interesting to compute the conditional VaR, which is the conditional expectation of R_p, given that it exceeds some constant, such as VaR_0. Note that if Z has the standard Normal distribution, $\mathcal{E}[Z \mid Z > z_0] = \phi(z_0)/[1 - \Phi(z_0)]$, where $\phi(z)$ is the probability density function and $\Phi(z)$ is the cumulative distribution function. Now, because R_p is distributed as $\mu_{r_p} + \sigma_{R_p} Z$, it follows that

$$\mathcal{E}[R_p \mid R_p > c] = \mu_{R_p} + \sigma_{R_p}\phi[(c - \mu_{R_p})/\sigma_{R_p}]/[1 - \Phi([(c - \mu_{R_P})/\sigma_{R_p}].$$

6.7 Selling Short

Here we have considered the weights a, b, c, etc., as positive. More generally, some of them could be allowed to be negative, representing short selling. We do not consider that here, leaving it for the reader's later exposure to the topic. A consideration is the necessary availability of resources to cover the short position in the event of a call on the stock.

6.8 Market Models and Beta

6.8.1 CAPM

Market models attempt to describe the RORs of individual stocks in terms of characteristics of the market as a whole.

The CAPM (Capital Asset Pricing Model) attempts to describe the RORs of individual stocks in terms of that of a market index, such as the S&P500. The CAPM was discussed earlier as an example of simple linear regression, either through the origin or not. Now this will be reviewed and related to the problem of computation of covariances in portfolio allocation. The model is often simply in the form $\mathcal{E}[Y|x] = \beta x$. Alternatively, a constant can be included, giving $\mathcal{E}[Y|x] = \alpha + \beta x$. The response variable Y is the ROR of the individual stock, the explanatory variable, x, the ROR of the market, as indicated, say, by that of the S&P500. Often, excess RORs are used; Y is the ROR of the individual stock minus the risk-free rate, and x is the ROR of the market minus the risk-free rate. To include time explicitly in the notation, write $\mathcal{E}[Y_t | x_t] = \alpha + \beta x_t$, $t = 1, 2, \ldots, n$. Often, n is taken to be 60, for sixty months (five years) of monthly data. For stocks $i = 1, 2, \ldots, m$, write $\mathcal{E}[Y_{it} | x_t] = \alpha_i + \beta_i x_t$. The constant α_i is the *differential return* or *abnormal return* of stock i.

6.8.2 Computation of Covariances under the CAPM

Now, under this model,

$$\mathcal{C}[Y_{it}, Y_{jt}] = \mathcal{C}[\alpha_i + \beta_i x_t + \varepsilon_{it}, \alpha_j + \beta_j x_t + \varepsilon_{jt}] = \mathcal{C}[\beta_i x_t + \varepsilon_{it}, \beta_j x_t + \varepsilon_{jt}]$$

because α_i and α_j are constants and do not affect the covariance. Continuing with the computation,

$$
\begin{aligned}
\mathcal{C}[\beta_i x_t + \varepsilon_{it}, \beta_j x_t + \varepsilon_{jt}] &= \mathcal{C}[\beta_i x_t, \beta_j x_t] + \mathcal{C}[\beta_i x_t, \varepsilon_{jt}] \\
&\quad + \mathcal{C}[\varepsilon_{it}, \beta_j x_t] + \mathcal{C}[\varepsilon_{it}, \varepsilon_{jt}] \\
&= \beta_i \beta_j \mathcal{V}[x_t] + 0 + 0 + 0 \\
&= \beta_i \beta_j \sigma_x^2.
\end{aligned}
$$

Here an assumption that the errors are uncorrelated is used. Taking the covariance of RORs of assets i and j to be simply $\beta_i \beta_j \sigma_x^2$ simplifies the computation of optimal portfolios.

The original form of the model in the formulation of the CAPM is simply $\mathcal{E}[Y_{it} | x_t] = \beta_i x_t$, without the constant term. The result on the simplification of the covariance of course still holds.

The *partial correlation coefficient* between x and y, given t, denoted by $r_{xy.t}$ is defined as the ordinary correlation between the parts of x and y that are not explained by t; that is, between the residuals in the regression of x on t and the residuals in the regression of y on t. Let $\hat{x} = a_{xt} + b_{xt}$, $\hat{y} = a_{yt} + b_{yt}$ be the predicted values of x and y based on t. Let the residuals be $\tilde{x} = x - \hat{x}$, $\tilde{y} = y - \hat{y}$. Then

$$r_{xy.t} = r_{\tilde{x}\tilde{y}},$$

where here r_{uv} denotes the ordinary correlation coefficient of x and y. The partial correlation coefficient can be computed in terms of the three correlations r_{xy}, r_{xt}, r_{ty} as

$$r_{xy.t} = \frac{r_{xy} - r_{xt} r_{ty}}{\sqrt{1 - r_{xt}^2} \sqrt{1 - r_{ty}^2}}.$$

To apply partial correlation in the Market Model approach to financial investments analysis, take t to be the market ROR, and x and y the returns of any two particular stocks of interest.

6.8.3   Section Exercises

6.1 Partial correlation. What is the value of $r_{xy.t}$ if $r_{xy} = .9, r_{xt,} = .6$, and $r_{ty} = .8$?

Solution. $[.9 - (.6)(.8)]/[(1 - .62)(1 - .82)]1/2 = .42/[(.6)(.8)] = .42/.48 = 7/8$ or .875. In this example, $r_{xy.t}$ has about the same value as r_{xy}.

6.2 Another set of correlations. What is the partial correlation $r_{xy \cdot t}$ if $r_{xy} = .9, r_{xt} = .6$, and $r_{ty} = .6$?

6.3 Interpret the partial correlation when x and y are the RORs of two stocks and t is the market ROR.

6.4 A "dummy" variable. Suppose that in the regression of Y on X and D the model is

$$\mathcal{E}[Y|X, D] = 4 + 3X + 2D + 2X D,$$

where D is a (0,1) variable. What is the difference, $\mathcal{E}[Y|X, D = 1] - \mathcal{E}[Y|X, D = 0]$? Note that in an application to stock returns, X could be the market return and D a Bull market indicator.

Solution: $(\hat{Y}$ when $D = 1) - (\hat{Y}$ when $D = 0) = (4 + 3X + 2 + 2X) - (4 + 3X) = 2 + 2X$ or $2(1 + X)$.

6.9   Summary

6.9.1   Rate of Return

Given a price series $\{P_t\}$, for a stock, the stock's rate of return (ROR) at time t is $R_t = (P_t - P_{t-1})/P_{t-1}$.

The continuous ROR r_t is the difference of successive log prices, $r_t = p_t - p_{t-1}$.

The continuous ROR is $r_t \approx R_t$, this approximation being close if P_t/P_{t-1} is close to 1. Also, r_t is a lower bound for R_t, that is, $r_t < R_t$.

6.9.2   Bi-Criterion Analysis

The possible portfolios can be represented in a plot of mean ROR versus standard deviation of ROR.

The best combination for a given risk-free rate is given by the tangency line to the feasible set. It has maximum Sharpe ratio.

6.9.3 Market Models

Market models attempt to explain the prices and RORs of stocks in terms of those of aspects of the market as a whole.

The CAPM (Capital Asset Pricing Model) regresses the ROR of any given asset on that of the market as a whole, represented, for example, by the S&P500 Composite Stock Index.

The financial analyst's "beta" is the coefficient in the regression of RORs of any particular asset on that of the market as a whole.

The usual and customary way to estimate a *beta* is from five years of monthly data ($n = 60$ RORs). The RORs used here may be excess RORs, that is, the ROR minus the risk-free rate. This regression may be taken through the origin, or a constant α may be included in the model. The constant α is called "abnormal" or "differential" return.

The covariance between asset RORs simplifies to the product of their betas times the variance of the market ROR.

6.10 Chapter Exercises

6.5 ROR. If a stock's share price goes from $50.00 to $50.50 in one period, what is the ROR? What is the continuous ROR?

6.6 ROR. If a stock's share price goes down from $50.50 to $50.00 in one period, what is the ROR? What is the continuous ROR?

6.7 Comparing up and down ROR. If a stock's share price goes from $100.00 to $90.00, it went down10%. What ROR is needed to get it back from $90.00 to $100.00?

6.8 (continuation) Analyze the same down and up situation in terms of continuous ROR.

6.10.1 Exercises on Covariance and Correlation

6.9 Variance of two uncorrelated variables. If $\mathcal{V}[U] = 4$, $\mathcal{V}[V] = 9$, and U and V are uncorrelated, what is $\mathcal{V}[U + V]$? *Answer:* $4 + 9 = 13$.

6.10 Variance of two positively correlated variables. If $\mathcal{V}[X] = 4$, $\mathcal{V}[Y] = 9$, and $\rho_{xy} = = +.8$, what is $\mathcal{V}[X + Y]$? *Solution:* $4 + 9 + (2)(.8)(2)(3) = 13 + 9.6 = 22.6$. This is larger than the variance of X and larger than the variance of Y. The standard deviation is $\sqrt{22.6} \approx 4.75$.

6.11 Variance of two negatively correlated variables. If $V[x] = 4$, $V[Y] = 9$ and $\rho_{xy} = -.8$, what is $V[X + Y]$? *Solution:* $4 + 9 + (2)(-.8)(2)(3) = 13 - 9.6 = 3.4$. This is smaller than the variance of X and smaller than the variance of Y. The standard deviation is $\sqrt{3.4} \approx 1.84$.

6.12 Covariance in terms of correlation and standard deviations. If the standard deviation of X is 7, the standard deviation of Y is 2, and the correlation of X and Y is $-.6$, what is $C[X, Y]$?

Solution: $C[X, Y] = \text{Corr}[X, Y] \, \text{SD}[X] \, \text{SD}[Y] = (-0.6)(7)(2) = -8.4$.

6.10.2   Exercises on Portfolio ROR

6.13 Portfolio mean. Given $\mu_x = 0.5\%$ per month and $\mu_y = 0.7\%$ per month, find a and $b = 1 - a$ such that $\mu_p = 0.5\%$ per month.

6.14 Portfolio mean. Given $\mu_x = 0.5\%$ per month and $\mu_y = 0.7\%$ per month, find a and $b = 1 - a$ such that $\mu_p = 0.65\%$ per month.

6.15 Portfolio variance. In the formula for $V[U + V]$, take $U = aX$, and $V = bY$, noting that $V[aX] = a^2\sigma_x^2$ and $V[bY] = b^2\sigma_y^2$. Then, for a portfolio with a weight of .1 on Stock A and .9 on Stock B, the portfolio variance is $V[.1X + .9Y] = .1^2V[X] + .9^2V[Y] + 2(.1)(.9)\,\text{Corr}[X, Y]\,\text{SD}[X]\text{SD}[Y]$. What is the numerical value of this expression if $\text{Corr}[X, Y] = -.6$?

Solution: The variance of the portfolio ROR is $(.01)(49) + (.81)(4) + (2)(.1)(.9)(-0.6)(7)(2) = 2.218$. *Remarks.* (i) The standard deviation is the square root of this, or about 1.489. (ii) The expected ROR of the portfolio is $(.1)(14)+(.9)(6) = 6.8\%$. To evaluate the loss probability $\Pr(\text{portfolio return} < 0)$, we have $z = (0 - 6.8)/\sqrt{2.218} = -6.8/1.489 = -4.566$. So $\Pr(\text{portfolio return} < 0) = \Pr\{Z < -4.566\} = \Pr\{Z > 4.566\}$. It can be shown that, for large values z_0, such as the 4.566 in this case, $\Pr\{Z > z_0\}$ is approximately $\phi(z_0) / z_0$, where $\phi(z)$ denotes the Normal p.d.f., $(1/\sqrt{2\pi})\exp(-z^2/2)$. For $z = z_0 = 4.566$, this is about $(0.3989)(.000030)/4.566 = .0000026$, or about three chances in a million. [Here $\exp(x)$ means e^x.] The loss probability of the portfolio is much smaller than that of Stock A alone or Stock B alone. (However, it should be pointed out that even when a Normal distribution fits most of a distribution, it does not necessarily fit well in the tails.)

6.16 Portfolio variance in terms of portfolio mean. Given $\mu_x = 10\%/\text{yr.}$, $\mu_y = 5\%/\text{yr.}$, $\sigma_x = 20\%/\text{yr.}$, $\sigma_y = 5\%/\text{yr.}$, and $\rho_{xy} = -0.5$, find A, B, and C in $\sigma_p^2 = A\mu_p^2 + B\mu_p + C$.

6.17 Portfolio variance in terms of portfolio mean. Given $\mu_x = 12\%/\text{yr.}$, $\mu_y = 6\%/\text{yr.}$, $\sigma_x = 20\%/\text{yr.}$, $\sigma_y = 3\%/\text{yr.}$, and $\rho_{xy} = -0.6$, find A, B, and C in $\sigma_p^2 = A\mu_p^2 + B\mu_p + C$.

6.18 Two stocks. Consider two stocks, A, which has a good mean ROR but is somewhat risky, and B, which has a moderate return but is less risky. The characteristics of their RORs x and y are given in Table 6.6. They are somewhat negatively correlated.

a. If the variable X has a Normal distribution, what is the probability of loss with Stock A alone, that is, what is $\Pr\{X < 0\}$? *Hint:* The standardized value of $x = 0$ is $z = (0 - 0.4)/0.8 = -0.5$.

[b.] What is the probability of loss with Stock B alone; that is, assuming Normality, what is $\Pr\{Y < 0\}$?

c. With weights $a = .6$ and $b = .4$, what is the portfolio mean ROR?

item[d.] (continuation) What is the variance of portfolio ROR?

e. What is the probability that the ROR of the portfolio will be negative?

TABLE 6.6
Two Stocks. Monthly RORs

Stock	A	B
ROR	x	y
Mean ROR	0.4 %	1.5 %
Std.Dev. of ROR	0.8 %	4.5 %
Corr. of RORs		$-.3$

6.19 Two other stocks. Consider two stocks, C, which has a good return but is somewhat risky, and D, which has a moderate return but is less risky (Table 6.7).

a. **Probability of loss with Stock C alone.** Assuming Normality, $\Pr\{X < 0\} = ?$

Solution: $z = (0 - 14)/7 = -2.00$, so the probability is about .0228. There is not much chance of a loss with Stock C.

b. Probability of loss with Stock D alone. Assuming Normality, $\Pr\{(Y < 0\} = ?$

Solution: $z = (0 - 6)/2 = -3.00$, prob $= .00135$. So there is even less chance of a loss with Stock D than with Stock C.

c. With weights $a = .6$ and $b = .4$, what is the portfolio mean ROR?

d. (continuation) What is the variance of portfolio ROR?

e. What is the probability that the ROR of the portfolio will be negative?

TABLE 6.7
Two Stocks. Annual RORs.

Stock	C	D
ROR	x	y
Mean ROR	14%	6%
Std.Dev. of ROR	7%	2%
Corr. of RORs		$-.6$

6.20 Variance of a sum of two uncorrelated variables. If $V[U] = 9$, $V[V] = 16$, and $\rho_{uv} = 0$, what is $V[U + V]$? What is $\mathrm{SD}[U + V]$?
Solution: The variance is $9 + 16 = 25$. The standard deviation is $\sqrt{25} = 5$.

6.21 Variance of a sum of two uncorrelated variables. If $V[U] = 25$, $V[V] = 144$, and $\rho_{uv} = 0$, what is $V[U + V]$? What is $\mathrm{SD}[U + V]$?
Solution: The variance is $25 + 144 = 169$. The standard deviation is $\sqrt{169} = 13$.

6.22 Variance of a sum of two positively correlated variables. If $V[X] = 4$, $V[Y] = 9$ and $\mathrm{Corr}[X, Y] = +.8$, what is $V[X + Y]$?
Solution: $4 + 9 + (2)(.8)(2)(3) = 13 + 9.6 = 22.6$, answer (E). Note that the standard deviation is $\sqrt{22.6} \approx 4.75$.

6.23 Covariance in terms of correlation and standard deviations. Continuing with the same X and Y, what is $C[X, Y]$?
Solution: $C[X, Y] = \rho_{xy}\sigma_x\sigma_y = (-0.6)(7)(2) = -8.4$.

6.24 Portfolio variance. The variance is $\mathcal{V}[(.1X + .9Y] = .1^2\mathcal{V}[X] + .9^2\mathcal{V}[Y] + 2(.1)(.9)\operatorname{Corr}[X, Y]\operatorname{SD}[X]\operatorname{SD}[Y]$. What is the numerical value of this?

Solution: The variance of the portfolio's ROR is $\mathcal{V}[R_p] = (.1^2)(7^2) + (.9^2)(2^2) + 2(.1)(.9)\sigma_{xy}$, where $\sigma_{xy} = \rho_{xy}\sigma_x\sigma_y = (-.6)(7)(2) = -8.4$, so $\mathcal{V}[R_p] = (.01)(49) + (.81)(4) + (2)(.1)(.9)(-8.4) = 2.218$. The standard deviation is the square root of this, or about 1.489.

Note. The expected return of the portfolio is $(.1)(14)+(.9)(6) = 6.8\%$. To evaluate $\Pr(\text{portfolio return} < 0)$, we have $z = (0 - 6.8)/\sqrt{2.218} = -6.8/1.489 = -4.566$. So $\Pr(\text{portfolio return} < 0) = \Pr\{Z < -4.566\} = \Pr\{Z > 4.566\}$. It can be shown that

$$\Pr\{Z > z_0\} \approx \frac{\phi(z_0)}{z_0}, \text{ for large } z_0,$$

where $\phi(z)$ is the expression for the bell-shaped Normal curve, $\phi(z) = (1/\sqrt{2\pi})\exp(-z^2/2)$. Values of z_0 greater than 4 could be considered large. In this case, $z_0 = 4.566$, so we have a right-tail probability of about $(1/\sqrt{2\pi})\phi(4.566)/4.566 \approx (0.3989)(.000030)/4.566 = .0000026 = 2.6 \times 10^{-6}$, or about three chances in a million. The loss probability of the portfolio is much smaller than that of Stock A or Stock B alone.

6.25 Minimum variance for two stocks. Given two stocks with RORs X and Y, show that the minimum variance weight a^* is given by

$$a^* = \mathcal{C}[Y, Y - X]/\mathcal{V}[Y - X].$$

6.26 (continuation) Show that this is the coefficient in the regression of Y on $Y - X$.

6.27 Find the minimum-variance weights a^*, b^* using a Lagrange multiplier. *Hint:* Define a Lagrangian function $L(a, b; \lambda) = [a^2\sigma_x^2 + b^2\sigma_y^2 + 2ab\sigma_{xy}] + 2\lambda(1 - a - b)$; compute the partial derivatives with respect to a, b, and λ; set them equal to zero; and solve.

6.28 Normal tail probability approximation. Use the right-tail approximation for $z_0 = 3$ and compare the approximation with the actual right-tail probability, which is about .00135.

6.29 Use the right-tail approximation for $z_0 = 3.5$ and compare the approximation with the actual right-tail probability, which is about .00023.

6.30 Show that $\mathcal{E}[e^{-aZ}] = e^{a^2/2}$ if Z has the standard Normal distribution.

6.31 Show that $\mathcal{E}[e^{-aR}] = e^{-a\mu + \sigma^2 a^2/2} = e^{-a(\mu - 1/2\,a\sigma^2)}$ if R has a Normal distribution with mean μ and variance σ^2.

6.10.3 Exercises on Three Stocks

6.32 Given $\mu_x, \mu_y, \mu_z, \sigma_x, \sigma_y, \sigma_z, \rho_{xz}, \rho_{xz}, \rho_{yz}$, find weights a, b, c ($a + b + c = 1$) such that the portfolio mean and standard deviation are equal to given numbers μ_p and σ_p. (Note that there are two free unknowns and two equations.)

6.33 Some triplets of correlations are not possible. In particular, the determinant of the correlation matrix must be positive. Verify that this is the case for the three correlations in Table 6.4.

6.34 Check to see whether the determinant of the correlation matrix formed from $r_{xy} = +.2, r_{xz} = +.3$, and $r_{yz} = -.9$ is positive.

6.35 Given $r_{xy} = +.2$ and $r_{xz} = +.3$, what range of values is possible for r_{yz}?

6.36 Check to see whether the determinant of the correlation matrix formed from $r_{xy} = +.2, r_{xz} = +.3$, and $r_{yz} = +.4$ is positive.

6.37 Given $r_{xy} = +.8$ and $r_{xz} = +.8$, what range of values is possible for r_{yz}?

6.10.4 Exercises on Correlation and Regression

6.38 The Market Model: "Beta". If a stock's beta is 1.10 and its alpha is 0.25% per month, what is its predicted return for a month in which the S&P500 goes up 0.3%? *Solution:* $0.25 + (1.10)(0.3) = 0.25 + 0.33 = 0.58\%$ per month.

6.39 If a stock's beta is 1.50 and its alpha is 0.25% per month, what is its predicted return for a month in which the S&P500 goes up 0.3%?

6.40 A stock with a small beta. If another stock's beta is 0.5 and its alpha is 0.40% per month, what is its predicted ROR in a month in which the S&P500 index goes up 0.3%? *Solution:* $0.40 + (0.5)(0.3) = 0.4 + 0.15 = 0.55\%$ per month.

6.41 If another stock's beta is 0.4 and its alpha is 0.20% per month, what is its predicted ROR in a month in which the S&P500 index goes up 0.3%?

6.42 If the correlation of Stock A's ROR with that of the S&P500 is .8, and the correlation of Stock B's ROR with that of the S&P500 is .7, what is the range of possibilities for the correlation of the RORs of Stock A and Stock B?

6.43 If the correlation of Stock C's ROR with that of the S&P500 is 8, and the correlation of Stock D's ROR with that of the S&P500 is .6, what is the range of possibilities for the correlation of the RORs of Stock C and Stock D?

6.44 If a stock's beta is 1.2 and the standard deviation of its ROR is 0.5% per month and the standard deviation of the S&P500 ROR is 0.3% per month, what is the correlation of the stock's ROR and that of the S&P500?

6.45 If a stock's beta is 0.8 and the standard deviation of its ROR is 0.5% per month and the standard deviation of the S&P500 ROR is 0.3% per month, what is the correlation of the stock's ROR and that of the S&P500?

6.11 Appendix 6A: Some Results in Terms of Vectors and Matrices (Optional)*

For those who may have used vectors and matrices before, some results are restated in those terms. *A bit of advice:* Although this section is marked as optional, it is highly recommended that students of mathematical / statistical finance become familiar with vector and matrix operations; such knowledge is a big help in shortening presentations and proofs, and, in my opinion, aids understanding by eliminating cumbersome scalar notation.

Vectors are denoted here by boldface lower-case letters; matrices, by boldface upper-case letters.

6.11.1 Variates

"Multivariate" statistical analysis' could instead be called "multivariable" statistical analysis. But it is not. Perhaps one reason that "variate" has been included in the name is that so much of the analysis devolves upon linear combinations ("variates").

Consider a variate in, say, m variables. That is, let

$$V = b_1 X_1 + b_2 X_2 + \cdots + b_m X_m.$$

Then, because the scalar product of two vectors is the sum of products of corresponding elements, V can be written as $V = \boldsymbol{b}' \boldsymbol{X}$, where, given a vector $\boldsymbol{v}$, the symbol $\boldsymbol{v}'$ denotes its transpose, $(v_1 \, v_2 \ldots v_m)$. If $\boldsymbol{v}$ is a column vector, then $\boldsymbol{v}'$ is a row vector. So, $\boldsymbol{X} = (x_1 \, x_2 \ldots x_m)'$. The **mean of a variate** V is $\mathcal{E}[V] = \boldsymbol{b}' \boldsymbol{\mu}$, where $\boldsymbol{\mu}$ is the mean vector of the random vector $\boldsymbol{X}$. The **variance of a variate** $V = \boldsymbol{b}' \boldsymbol{X}$ is $\mathcal{V}[V] = \mathcal{V}[\boldsymbol{b}' \boldsymbol{X}] = \boldsymbol{b}' \boldsymbol{\Sigma} \boldsymbol{b}$, where $\boldsymbol{\Sigma}\,(m \times m)$ is the covariance matrix of the random vector $\boldsymbol{X}$. The *covariance of two variates* $\boldsymbol{a}' \boldsymbol{X}$ and $\boldsymbol{b}' \boldsymbol{X}$ is $\mathcal{C}[\boldsymbol{a}' \boldsymbol{X}, \boldsymbol{b}' \boldsymbol{X}] = \boldsymbol{a}' \boldsymbol{\Sigma} \boldsymbol{b}$.

6.11.2 Vector Differentiation

Given a function $f(x)$ where $x = (x_1 \, x_2 \, \ldots \, x_m)'$, the partial derivative of f with respect to x_j is denoted by $\partial f / \partial x_j$. The vector of partial derivatives is

$$\partial f / \partial x = (\partial f / \partial x_1 \; \partial f / \partial x_2 \; \ldots \; \partial f / \partial x_m)'.$$

This vector is known as the *gradient*. (See also Appendix A.)

6.11.2.1 Some Rules for Vector Differentiation

It is easy to verify various rules for vector differentiation, such as the following. Analogous to

$$f'(ax) = \frac{d \, ax}{dx} = a,$$

we have

$$\frac{d \, ax}{dx} = a \, 1,$$

where 1 is the vector of ones.

Analogous to

$$f'(ax^2) = \frac{d \, ax^2}{dx} = 2ax,$$

we have

$$\frac{d \, x' A x}{dx} = 2 \, A \, x.$$

6.11.2.2 Minimum-Variance Portfolio

In terms of the weight vector w and covariance matrix S, the portfolio variance is $w' S w$. This is to be minimized with respect to w, subject to constraints.

Including the Lagrange multiplier for the constraint $1'w = 1$, the function V to be minimized is $V(w; \lambda) = w' S w + 2\lambda(1 - 1'w)$. (It is convenient to enter the Lagrange multiplier as 2λ rather than λ.) The vector derivative with respect to w is $\partial V / \partial w = 2 S w - 2\lambda 1$. Setting this equal to zero gives $S w = \lambda 1$. Now let T denote the inverse of S. Pre-multiplying by T gives $w = \lambda T 1$. Now pre-multiplying by $1'$ gives $\lambda 1'T 1 = 1'w = 1$, and $\lambda = 1/1'T 1$. This gives the minimum-variance weight vector w^* as $w^* = T \, 1/1'T 1$. Thus the optimal weight vector w^* is in the direction of $T 1$. That is, $w* = \text{Const.}\, T 1$, where the constant is determined by the condition that the elements of w^* sum to 1. This solution allows weights to be negative. When the constraint of positive weights is imposed, mathematical programming (quadratic programming) must be used to get the solution.

Next the maximization of the Sharpe ratio is considered.

6.11.2.3   Maximum Sharpe Ratio

Given weight vector w, mean vector m, and covariance matrix S, the portfolio mean is $m'w$ and the portfolio variance is $w'Sw$.

The Sharpe ratio S is $S = (\mu_p - r_f)/\sigma_p = (m'w - r_f)/\sqrt{w'Sw}$. Here this is to be considered as a function $S(w)$ of w.

The resulting optimal value of w, say w^*, is in the direction $T\,m$, where the matrix T is the inverse of the matrix S. That is, $w^* = \text{Const. } T\,m$, where   Const. is determined by the condition that the elements of w^* sum to 1.

To derive the optimal w^*, note that the function to be maximized, including the Lagrange multiplier for the constraint, $1'w = 1$, is the Lagrangian

$$L(w;\lambda) = S(w) + \lambda(1 - 1'w) = m'w/w'Sw + \lambda(1 - 1'w).$$

Now, $S(w) = N(w)/D(w)$, where the numerator $N = m'\,m - R_f$ and the denominator $D = (w'Sw)^{1/2}$. The gradient of S with respect to w is $dS = (dN \cdot D - N \cdot dD)/D^2 = $. Using this, taking partials of L with respect to w and λ, setting them equal to zero and solving gives, after some algebra, $w^* = C(w^*)S^{-1}m$, where C is a scalar. Although C depends upon w^*, the fact that it is a scalar means that w^* is in the direction given by the vector $S^{-1}m$. The scalar C is determined by the condition that the elements of the weight vector sum to one.

This solution allows weights to be negative. When the constraint of positive weights is imposed, mathematical programming (quadratic programming) must be used to get the solution.

6.11.3   Section Exercises

6.46 Find the minimum variance weight vector $w^* = (a^* \ b^*)'$ for two stocks whose RORs have standard deviations 0.8% and 0.3% per month and a correlation of $-.4$.

6.47 Find the minimum variance weight vector $w^* = (a^* \ b^*)'$ for two stocks whose RORs have standard deviations 0.8% and 0.3% per month and a correlation of $+.4$.

6.48 Denoting the RORs of $m = 2$ stocks by x and y, verify that the elements of the inverse S^{-1} are $s^{(xx)} = s_{yy}/D$, $s^{(yy)} = s_{xx}/D$, $s^{(xy)} = s^{(yx)} = s_{xy}/D$, where D is the determinant, $|S| = s_{xx}s_{yy} - s_{xy}^2$.

6.49 (continuation) Show that $D = s_{xx}s_{yy}(1 - r_{xy}^2)$.

6.50 (continuation) Show that

$$s^{(xx)} = \frac{1}{s_{xx}(1 - r_{xy}^2)}, \quad s^{(yy)} = \frac{1}{s_{yy}(1 - r_{xy}^2)}, \quad s^{(xy)} = s^{(yx)} = \frac{-r_{xy}}{s_x s_y(1 - r_{xy}^2)},$$

where s_x, s_y are the standard deviations of x and y. That is, show that the inverse covariance matrix is as follows.

$$S^{-1} = \frac{1}{1 - r_{xy}^2} \begin{bmatrix} 1/s_{xx} & -r/s_x s_y \\ -r/s_x s_y & 1/s_{yy} \end{bmatrix}$$

6.51 In Section 6.2 on two stocks, the minimum variance weights were derived for that case. Verify that the general result in terms of vectors and matrices for m stocks reduces to the result obtained there.

6.12 Appendix 6B: Some Results for the Family of Normal Distributions

(See also the appendices at the end of the book .)
The family of Normal distributions is indexed by the parameter pair (μ, σ^2), the mean and variance. The probability density function of the Normal distribution with mean μ and standard deviation σ is

$$\phi(x; \mu, \sigma^2) = \frac{1}{\sqrt{2\pi}} \exp[-\frac{1}{2\sigma^2}(x - \mu)^2], \quad -\infty < x < \infty.$$

If Z has the standard Normal distribution, then its probability density function is

$$\phi(z) = \frac{1}{\sqrt{2\pi}} \exp(-\frac{1}{2}z^2), \quad -\infty < z < \infty.$$

6.12.1 Moment Generating Function; Moments

If Z has this distribution, then $\mathcal{E}[e^{tX}]$, the moment generating function, is $\mathcal{E}[e^{tX}] = \exp(1/2\,t^2)$. To see this, note that

$$\mathcal{E}[e^{tZ}] = \int_{-\infty}^{\infty} \exp(tz)\,\phi(z)\,dz = \int_{-\infty}^{\infty} \exp(tz)\,(1/\sqrt{2\pi})\exp[(-1/2)z^2]\,dz.$$

Combine the exponents of e and complete the square to obtain the result. If X has a Normal distribution with mean μ and variance σ^2, then X is distributed as $\mu + \sigma Z$. It follows then that $\mathcal{E}[e^{tX}] = \exp(t\mu + 1/2\,t^2\sigma^2)$.

6.12.2 Section Exercises

6.52 Use the preceding result to show that if X has a Normal distribution, then, for any real number s,

$$\mathcal{E}[-s\,X] = \exp(-s\mu + 1/2s^2\sigma^2) = \exp[(-s)(\mu - 1/2s\sigma^2)].$$

Remarks. (i) This result will be used in the next chapter. (ii) If $-s$ is replaced by t, then the resulting function is called the *moment generating function.*

6.53 Compute $\mathcal{E}[\,Z\,|\,Z > z_0]$ if Z has the standard Normal distribution. *Hints:* The conditional probability density function is $\phi(z)/[1 - \Phi(z_0)]$. Perform the indicated integration using the substitution $u = z^2/2$, $du = z\,dz$.

6.13 Bibliography

Bodie, Zvi, Kane, Alex, and Marcus, Alan (2009). *Investments. 8th ed.* McGraw-Hill / Irwin, New York.

Markowitz, Harry M. (1959). *Portfolio Selection: Efficient Diversification of Investments.* John Wiley & Sons, New York. (Reprinted by Yale University Press, 1970, 2nd ed. Basil Blackwell, 1991.)

Ross, Sheldon M. (2003). *An Elementary Introduction to Mathematical Finance. 2nd ed.* Cambridge University Press, New York and Cambridge.

Ross, Sheldon M. (2010). *Introduction to Probability Models. 10th ed.* Academic Press (Elsevier), Burlington, MA.

Ruppert, David (2011). *Statistics and Data Analysis for Financial Engineering.* Springer, New York.

Wagner, Niklas (2002). On a model of portfolio selection with benchmark. *Journal of Asset Management,* **3,** 55–65.

6.14 Further Reading

The book by Bodie, Kane, and Marcus (2009) is a new edition of a very widely used introduction to financial investments analysis. Ross (2003) is a particularly clear and helpful introduction to mathematical finance. Ruppert (2011) is an excellent graduate-level textbook on statistical finance.

Among articles, Wagner (2002) is especially recommended.

7

Utility-Based Portfolio Analysis

CONTENTS

7.1   Introduction

This chapter and the preceding one form a unit on portfolio analysis. In this chapter, the notion of single-criterion portfolio analysis is continued from the preceding chapter. Here, a criterion that combines portfolio mean and variance is derived.

7.1.1   Background

To review, recall that *bi-criterion* portfolio analysis, or *mean-variance* analysis, is based on the mean and variance of portfolio ROR. A good portfolio is one with a good combination of portfolio mean ROR μ_p and variance σ_p^2 of portfolio ROR, namely, relatively high mean and relatively low variance. The possible combinations are usually represented in a plot of mean versus standard deviation or mean versus variance.

7.1.2   Types of Portfolio Analysis

Single-criterion analysis combines the mean μ_p and variance σ_p^2 of the port-folio into a single function, such as $\mu_p - 1/2\,A\,\sigma_p^2$, for a suitable value of the *risk-aversion constant* A. This function is referred to as a *functional*, meaning in this case that it is a function of a distribution, namely, the distribution of R_p, through its mean μ_p and variance σ_p^2. This functional is called the *utility-based functional* or simply *utility functional*, because, as will be seen below, it is derived from a utility function. (The factor $1/2$ that appears results from this derivation.)

The constant A varies from investor to investor. The expression $\mu_p - 1/2\,A\,\sigma_p^2$ may be viewed as penalized ROR, where the penalty term $1/2\,A\,\sigma_p^2$ involving the variance is subtracted from the mean ROR μ_p. The scale factor $\frac{1}{2}$ (in $A/2$) emerges from the analysis to be used here, based on an exponential utility function for portfolio ROR, so we retain this factor. Details are given later in the chapter.

To review, remember that, in general, combination portfolios are considered, meaning the combination of a risk-free portion and a risky portion. The ROR of the combination portfolio is $R_p = w_f R_f + w_r R_r$, where R_r denotes the ROR of the risky part and R_f is the risk-free rate. (The subscript r in w_r and R_r denotes "risky.") Note that

$$R_p = w_f R_f + w_r R_r = (1 - w_r)R_f + w_r R_r = R_f + w_r(R_r - R_f).$$

Next, review the expressions for the mean and variance of portfolio ROR in terms of the mean μ_r and variance σ_r^2 of the risky portion. The expected value of portfolio ROR is

$$\begin{aligned}
\mathcal{E}[R_p] &= \mathcal{E}[w_f R_f + w_r R_r] \\
&= w_f R_f + w_r \mu_r \\
&= (1 - w_r)R_f + w_r \mu_r \\
&= R_f + w_r(\mu_r - R_f).
\end{aligned}$$

The variance of portfolio ROR is, remembering that R_f is taken as fixed, not random,

$$\mathcal{V}[R_p] = \mathcal{V}[w_f R_f + w_r R_r] = \mathcal{V}[w_r R_r] = w_r^2 \sigma_r^2.$$

The choice of values of w_f and w_r separates from the problem of allocation to assets in the risky portion. Next we illustrate this in the context of single-criterion analysis.

7.2   Single-Criterion Analysis

A single criterion incorporates both mean and variability. Such a criterion is $\mu_{R_p} - 1/2\,A\,\sigma_{R_p}^2$.

This criterion arises from an *exponential utility function* for wealth (the investor's "fortune," or at least that part of it devoted to investment). (See, for example, Ross 2011.) Denote wealth by s. An exponential utility function for wealth is

$$U(s) = 1 - \exp(-a\,s),\ s > 0,$$

where $a > 0$ is a constant that varies from investor to investor.

What does this utility function for wealth imply about that of next period's portfolio ROR, R_{pt}? Now, $s_t = s_{t-1}(1 + R_{pt})$, where s_{t-1} is the wealth at the end of the previous period and s_t is that at the end of period t. Thus,

$$U(s_t) = 1 - \exp(-a\,s_t) = 1 - \exp[-a\,s_{t-1}(1 + R_{pt})].$$

Next, the conditional expected value of utility, given s_{t-1}, is computed. This proceeds as

$$
\begin{aligned}
\mathcal{E}[\exp(-a\,s_t) \mid s_{t-1}] &= \mathcal{E}[\exp[-as_{t-1}(1 + R_{pt})] \mid s_{t-1}] \\
&= \exp(-as_{t-1})\,\mathcal{E}[\exp(-as_{t-1}\,R_{pt}) \mid s_{t-1}] \\
&= \exp(-A)\,\mathcal{E}[\exp(-AR_{pt}) \mid s_{t-1}],
\end{aligned}
$$

where $A = as_{t-1}$. Remember, however, that this A depends upon the level s_{t-1} of wealth at time $t-1$.

Now this is evaluated under an assumption that R_{pt} has a Normal distribution. If X is distributed according to a Normal distribution with mean μ and variance σ^2, then

$$\mathcal{E}[\exp(-AX)] = \exp(-A\mu + 1/2\,A^2\sigma^2) = \exp[(-A)(\mu - 1/2\,A\,\sigma^2)].$$

This gives

$$
\begin{aligned}
\mathcal{E}[U(s_t)] &= 1 - \exp(-A)\,\mathcal{E}[\exp(-AR_{pt}) \mid s_{t-1}] \\
&= 1 - \exp(-A)\mathcal{E}[\exp[(-A)(\mu - 1/2\,A\sigma^2)].
\end{aligned}
$$

The expected utility is maximized by maximizing $\mu - 1/2\,A\,\sigma^2$ over combinations (μ, σ^2). Remember that in the present application, $\mu = \mu_p$ and $\sigma^2 = \sigma_p^2$, the mean and variance of the portfolio ROR, respectively. That is, the functional to be maximized is $\mu_p - 1/2\,A\,\sigma_p^2$. Here this criterion is called the *utility-derived* or *utility-based* functional.

This particular functional was derived assuming Normality of portfolio ROR, but it is perhaps not unreasonable more generally (although we must admit that mean ROR minus a multiple of standard deviation—rather than variance—seems more natural in a way).

Remember that here $A > 0$ is a given constant that varies from investor to investor and depends on the investor's level of aversion toward risk and level of wealth. The value of the risk-aversion constant A varies with the preferences of the individual investor and A and the investor's level of wealth and might be between, say, 2 and 20 or even 100. See, for example, Bodie, Kane, and Marcus (2009), who work examples (p. 159) with A equal to 2, 3.5, and 5.

The criterion in the form $\mu_p - 1/2 A \sigma_p^2$ is for μ_p and σ_{R_p} given as decimals. To use percents, note that this is $\mu_{R_p}\%/100 - 1/2 A (\sigma_p\%/100)^2 = 0.01 [\mu_p\% - 1/2 A (0.01)(\sigma_p\%)^2]$. Ignoring the constant of (0.01) in the front, because it does not affect the comparison of portfolios, one can work in terms of

$$\mu_p\% - (0.01)\,1/2\,A\,(\sigma_p\%)^2.$$

That is, with RORs in percent, the criterion is expressed as

$$\mu_p - (1/2)(0.01)\,A\sigma_p^2, \text{ or } \mu_p - 0.005\,A\sigma_p^2.$$

In general, utility does not have units, but here, with μ in percent, the units can be taken as percent. Values of utility for various values of μ, σ, and A are given in Table 7.1.

TABLE 7.1
Utility for Various μ, σ, A

μ	σ	$A = 1$	$A = 2$	$A = 4$	$A = 10$
0.1	0.05	0.0999875	0.099975	0.09995	0.099875
	0.10	0.09995	0.0999	0.0998	0.0995
	0.20	0.0998	0.0996	0.0992	0.098
1	0.50	0.99875	0.9975	0.995	0.9875
	1.00	0.995	0.99	0.98	0.95
	2.00	0.98	0.96	0.92	0.8
2	1.00	1.995	1.99	1.98	1.95
	2.00	1.98	1.96	1.92	1.8
	4.00	1.92	1.84	1.68	1.2

Examples. (i) Suppose the standard deviation of monthly ROR is 1% but increases from 1% to 2%. What increase in monthly ROR would be necessary to compensate for that? Because the standard deviation σ increases from 1% to 2%, the variance σ^2 increases from 1 to 4, so the term $1/2A\sigma_p^2$ increases from $1/2\,(0.01)A(1)$ to $1/2\,(0.01)\,A\,(4)$, an increase of $1/2\,A(0.01)(3) = 0.015A$. The compensating increase in mean monthly ROR would be $0.015\,A\%$. If $A = 4$, this is 0.06 %. If $A = 100$, this is 1.5%.

(ii) Now suppose the standard deviation of monthly ROR was 1% but increased from 1% to 3%. What increase in monthly ROR would be necessary to compensate for that? Because the standard deviation σ increases from 1% to 3%, the variance σ^2 increases from 1 to 9, so the term $1/2A\sigma_p^2$ increases from $1/2\,(0.01)A(1)$ to $1/2\,(0.01)\,A\,(9)$, an increase of $1/2\,A(0.01)(8) = 0.04A$. The compensating increase in mean monthly ROR would be $0.04\,A\%$. If $A = 4$, this is $0.16\ \%$. If $A = 25$, it is $1.0\ \%$, a 1% increase in mean ROR compensating for a 2% increase in ROR standard deviation. If $A = 100$, it is 4.0%, a 4% increase in mean ROR compensating for a 2% increase in ROR standard deviation.

7.2.1 Mean versus Variance Plot

Refer to the plot in Figure 7.1. This is a plot in the (σ^2, μ)-plane, that is, the plane with the horizontal (x) axis being σ^2 and the vertical (y) axis being μ, the locus of constant value c for $\mu - 1/2\,A\,\sigma^2$ is $\{(\sigma^2, \mu) : \mu - 1/2\,A\,\sigma^2 = c\}$. Written in the form $y = a + bx$, this is $\mu = c + 1/2\,A\,\sigma^2$. If $y = a + bx$, the slope is b, so here the slope is $A/2$. To maximize the criterion, start with a high value of c that gives a line above the set $\mathcal{F}$ of feasible pairs (σ^2, μ). As you decrease c, you generate parallel lines moving closer to $\mathcal{F}$. Move down until you get a line just touching $\mathcal{F}$; this line is the tangent line. The point (σ^{2*}, μ^*) of tangency gives the maximum value of the criterion $\mu - \frac{1}{2}\,A\sigma^2$.

7.2.2 Weights on the Risk-Free and Risky Parts of the Portfolio

Now consider further the case of a risk-free asset and m risky assets. The portfolio ROR is $R_p = w_f R_f + w_r R_r$, where here R_r is the ROR of the risky portion. The value of the utility-based functional is

$$\mu_{R_p} - \tfrac{1}{2}\,A\sigma^2_{R_p} \;=\; R_f + w_r(\mu_{R_r} - R_0) - \tfrac{1}{2}\,A\,w_r^2\sigma^2_{R_r}.$$

Optimal weights on the risk-free and risky parts can be calculated. Next we show this for the single criterion. The criterion is $R_f + w_r(\mu_r - R_f) - \frac{1}{2}\,A\,w_r^2\sigma_r^2 = ax^2 + bx + c$, with $x = w_r$, $a = -\frac{1}{2}\,A\,\sigma^2_{R_r}$, $b = \mu_{R_r} - R_f$, and $c = R_f$. Now, the optimal value of w_r is

$$w_r^* = x^* = -\frac{b}{2a} = -\frac{\mu_{R_r} - R_f}{(2)(-\frac{1}{2}\,A\,\sigma^2_{R_r})} = \frac{\mu_{R_r} - R_f}{A\,\sigma^2_{R_r}}.$$

This is increasing in μ_{R_r} and decreasing in R_f, A, and $\sigma^2_{R_r}$.

7.2.3 Separation

From this it is seen that the calculation of optimal weight on the risk-free asset separates from the computation of the optimal allocation within the risky portion of the portfolio. That is, given *any* allocation of weights in the risky portion, let μ_{R_r} and $\sigma^2_{R_r}$ denote the resulting mean and variance. Then the above calculation of w_r^* can be carried out.

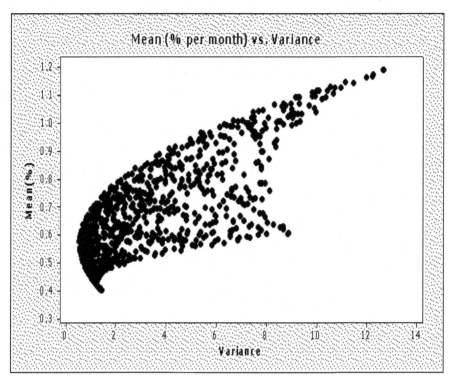

FIGURE 7.1
Mean versus variance

Even if the investor does not wish to rebalance the weights in the risky portion for the next time period, the allocation to the risk-free and risky portions could be changed appropriately according to updated estimates of the mean and variance of the risky portion's ROR.

7.3 Summary

The *utility functional* is $\mu_p - 1/2\,A\sigma_p^2$, where $A > 0$ is the investor's *risk aversion constant*.

This functional is derived from exponential utility for wealth. The constant A depends upon the investor's level of wealth.

An investor's total portfolio can be divided into a risk-free and a risky portion. The question of optimal allocation to the two parts is considered.

7.4 Chapter Exercises

7.1 Two stocks. Consider two stocks, A and B, with RORs x and y. Write out the utility functional in terms of their means μ_x, μ_y; standard deviations σ_x, σ_y; and correlation ρ_{xy}. Denote the weights by a and b.

7.2 An investor with risk-aversion constant $A = 20$ holds a portfolio with $\mu_p = 1.5\%$ per month and $\sigma_p = 1.2\%$ per month, and a risk-free rate of 0.1% per month. What are the optimal weights w_r^* and w_f^*?

7.3 An investor with risk-aversion constant $A = 10$ holds a portfolio with $\mu_p = 1.5\%$ per month and $\sigma_p = 1.2\%$ per month, and a risk-free rate of 0.1% per month. What are the optimal weights w_r^* and w_f^*?

7.4 Describe the locus of constant value of the utility functional in the (σ^2, μ)-plane.

7.5 Discuss estimation of the utility functional $\mu - 1/2\,A\,\sigma^2$, based on unbiased estimates m, the sample mean, and s^2, the sample variance, for μ and σ^2, respectively.

7.6 (continuation) If m and s^2 are based on a random sample of n from a Normal distribution, what is the variance of the estimator given in answer to the preceding problem? *Hints:* Given Normality, m and s^2 are statistically independent, hence uncorrelated. The sample mean m is distributed according to a Normal distribution with mean μ and variance σ^2/n. The r.v. $Q = (n-1)\,s^2\,/\,\sigma^2$ is distributed according to a chi-square distribution with $n-1$ d.f., so $\mathcal{E}[Q] = n - 1$ and $\mathcal{V}[Q] = 2(n-1)$.

7.5 Bibliography

Bodie, Zvi, Kane, Alex, and Marcus, Alan (2009). *Investments. 8th ed.* McGraw-Hill/Irwin, New York.

Ross, Sheldon M. (2011). *An Elementary Introduction to Mathematical Finance. 3rd ed.* Cambridge University Press, New York and Cambridge.

Ruppert, David (2011). *Statistics and Data Analysis for Financial Engineering.* Springer, New York.

Part IV

TIME SERIES ANALYSIS

8

Introduction to Time Series Analysis

CONTENTS

8.1   Introduction

This chapter concerns various models for time series data. A *time series* will be denoted by notation such as

$$\{Y_t,\ t = 1, 2, \ldots, n\}.$$

This notation refers to a variable Y observed at time points t.

Such data, observed in time, are *temporal data*. The observations may be triggered by events, occurring at irregular intervals. Or the observation times may be planned, in which case they can be at regular intervals.

Tick-by-tick data (*tick data*) on the price of a stock are temporal data observed at irregular intervals. These are triggered by a purchase or sale. These data report the prices at each transaction of a specific stock. The subscript t refers to the t-th transaction.

Often by the phrase "time series" people mean temporal data observed at regular time intervals, for example, daily, weekly, monthly, quarterly, or annually. The daily closing prices of a stock are a time series observed at regular intervals, namely, at the closing bell, marking the closing time of the exchange each trading day. Then the subscript t refers to the t-th day. For example, t might be in quarters (three-month periods) and Y might be quarterly sales of a retail company. Then Y_5, for example, represents the sales in the fifth quarter in the dataset.

A *time-series plot* is a plot of the variable Y against t.

8.2 Control Charts

Time-series plots are used in the control of manufacturing and service processes. The variable monitored may be, to mention a few diverse examples, the diameters of bolts, or the lengths of time to make deliveries to customers, or the concentration of salt in potato chips. (In regard to the latter, see Annenberg Foundation (1989), Program 18, "The Mean and Control Charts." This and other videos in that series are highly recommended.)

A *control chart* is a time series chart with upper and lower limits indicated. These limits may, for example, be at three standard deviations above and below a center line defined by the mean. Data points falling outside the limits can be investigated to determine if there is a particular cause (an "assignable cause") for the deviation. As these causes are detected and understood, corrective actions can be taken to prevent future results from going "out of control." Long periods above or below the center line indicate that a change or shift in the process has occurred. The point at which the shift occurred should be investigated, so corrective action can be taken to bring the process back into control. In statistical process control (SPC), a time-series plot is also called a *run chart*. In addition to several (say nine) observations on one side of the center line, other rules are used. Such rules are called *run signals, decision rules, multirules*, or simply *signals*. One such signal is the occurrence of two out of three points in a row between 2 and 3 standard deviations above the center line, or below the center line. The signals are defined to have false-alarm rates on the order of one in a thousand. For example, the probability of nine observations in a row above the median is $(1/2)^9$ or $1/512 \approx .002$. The probability of two of out three points in a row between 2 and 3 standard deviations above the center line in the Normal case is $3\,[\Phi(3) - \Phi(2)]^2 = 3(.99865 - .97725)^2 = 3(.02140)^2 \approx 3 \times 4.58 \times 10^{-4} \approx 1.37 \times 10^{-3} = .00137$. The reason why these tests are set to have Type I error rates near one in a thousand rather than the one in twenty used in conventional hypothesis testing is that the tests are done repeatedly, for each point of the chart, and the overall error rate needs to be controlled.

There are many current and potential applications of control charts, and not just in manufacturing. For example, in accounting, efficiency could be measured and charted. A variable to observe might be the number of days it takes to process an order, from receipt of an invoice to the time the order is shipped to the customer. Control charts can help detect errors in data, such as by charting the weekly payroll. A week where the payroll is significantly higher than prior weeks would be investigated to make sure there is a valid explanation. Travel and entertainment expenses can be similarly monitored.

Control charts may be based on individual observations. In this case, the chart is called an *I chart, I* for individual observations. Although such charts can be helpful, there is a problem: the individual observations may be correlated, so that the use of two- or three-sigma limits based on random samples is not valid. Methods of dealing with such correlation are discussed in this chapter.

Instead of individual observations, control charts may be based on *sliding* averages, for example the mean of the first three observations, then the mean of the next three, etc. That is, the window is for $t = 1, 2,$ and $3,$ then for $t = 4, 5, 6,$ etc. First the window reveals the observations for $t = 1, 2, 3,$ then it slides over to $t = 4, 5, 6.$ Control charts based on the mean are called $\bar{x}$-*charts,* "x-bar charts."

Control charts may also be based on *moving* averages, where the window is moved one point at a time, for example, for times 1, 2, 3, then for times 2, 3, 4, etc. A moving average is updated by dropping the oldest observation and adding in the newest. When control charts are based on sliding or moving averages, the effects of correlation are somewhat mitigated.

The idea of control charts and moving averages is applied in many fields, including financial investments analysis. Long-term and short-term moving averages are computed and compared; see in particular the average called MACD later in the chapter (Section 8.3.4.3).

8.3 Moving Averages

Moving averages may be used to smooth a series and thereby to spot trends. The average used may be a mean or a median.

8.3.1 Running Median

Another name for moving average is *running average.* "Running" refers to a moving window. A running average of three is a moving average formed with a window of width three. That means computing an average of observations 1, 2, and 3, then of observations 2, 3, 4, then of observations 3, 4, 5, etc.

For three observations, the median is the middle (second-ranking) observation. If the minimum and maximum of three observations are trimmed off, one observation remains, and that is the median of the three.

Given a series $\{y_t, t = 1, 2, \ldots, n\}$, the smoothed value, the running median of three, say rMdn_t, is

$$\text{rMdn}_t = \text{median}\{y_{t-1}, y_t, y_{t+1}\}, \ t = 2, 3, \ldots, n - 1.$$

Computing a moving average is one method of *smoothing* a series, and the values of the moving average are called "smoothed" values.

Example 8.1 Running median of three

Suppose the value of the S&P500 index on seven successive trading days was $y_1 = 1135, y_2 = 1137, y_3 = 1134, y_4 = 1129, y_5 = 1128, y_6 = 1127, y_7 = 1125$. The computation of the running median of three proceeds as

$$\text{rMdn}_2 = \text{median}\{1135, 1137, 1134\} = 1135,$$

$$\text{rMdn}_3 = \text{median}\{1137, 1134, 1129\} = 1134,$$

and so on, obtaining smoothed values $\text{rMdn}_4 = 1129, \text{rMdn}_5 = 1128$, and $\text{rMdn}_6 = 1127$. It will be noted that the plot of the rMdn sequence will be smoother than that of the original sequence. Note the consistent downtrend of the smoothed series, despite the slight ups and downs of the original series.

8.3.2 Various Moving Averages

A moving average with a window of width four is especially appropriate for quarterly data because then the smoothed results are balanced, in that each smoothed result contains each quarter once and only once.

A moving average may be *centered,* like the RM above, or may work back in time, like $\text{MA}_t = (y_t + y_{t-1} + y_{t-2} + y_{t-3})/4$. An example of a *weighted* four-period moving average that weights more recent observations more heavily is

$$\text{WMA}_t = .4y_t + .3y_{t-1} + .2y_{t-2} + .1y_{t-3}.$$

Note that $.1 + .2 + .3 + .4 = 1$. A five-period weighted MA is

$$(5y_t + 4y_{t-1} + 3y_{t-2} + 2y_{t-3} + y_{t-4})/15,$$

where we have divided by 15 because $1 + 2 + 3 + 4 + 5 = 15$. An n-period sum-of-digits weighted MA is

$$[ny_t + (n-1)y_{t-1} + \cdots + 2y_{t-n+2} + 1y_{t-n+1}]/[n(n+1)/2],$$

because $1 + 2 + 3 + \cdots + n = n(n+1)/2$.

8.3.3   Exponentially Weighted Moving Averages

An *exponentially weighted moving average* has coefficients that decrease exponentially as you go back in time; that is, the coefficients are the terms in a geometric series. An example of this is

$$.3\,y_t \; + \; (.3)(.7)\,y_{t-1} \; + \; (.3)(.7^2)\,y_{t-2} \; + \; (.3)(.7^3)\,y_{t-3}$$
$$= \; .3\,y_t \; + \; .21\,y_{t-1} \; + \; .147\,y_{t-2} + .1029\,y_{t-3}.$$

It can be divided by the sum of the weights so that the resulting weights sum to one. Recall that $1 + r + r^2 + \cdots = 1/(1-r)$ for $|r| < 1$, and

$$1 + r + r^2 + \cdots + r^n \; = \; (1 - r^{n+1})/(1-r), \text{ for } r \neq 0.$$

So in this example the sum of the weights is $.3(1 - .7^4)/(1 - .7) = 1 - .7^4 = .7599$; dividing by this gives

$$\text{EWMA}_t \; = \; (.3y_t + .21y_{t-1} + .147y_{t-2} + .1029y_{t-3})/.7599$$
$$\approx \; .395y_t + .276y_{t-1} + .193y_{t-2} + .135y_{t-3}.$$

Note the similarity to the WMA $.4y_t + .3y_{t-1} + .2y_{t-2} + .1y_{t-3}$.

In some applications, such as Statistical Quality Control, exponentially weighted moving averages are indeed called EWMAs, pronounced "You may." In financial investments analysis, these are usually called simply EMAs— Exponential Moving Averages.

An EWMA can be calculated in terms of the current observation and the preceding value of the EWMA as the weighted average

$$\text{EWMA}_t = \alpha y_t + (1 - \alpha)\text{EWMA}_{t-1},$$

where EWMA_t denotes the smoothed value and $\alpha\,(0 < \alpha < 1)$ is called the *smoothing constant*. The smoothing constant is the weight placed on the most recent observation. (It would seem more appropriate to refer to $1 - \alpha$ as the smoothing constant, as larger values of $1 - \alpha$, that is, smaller values of α, produce more smoothing.)

Note that

$$\text{EWMA}_t \; = \; \alpha y_t + (1 - \alpha)[\alpha y_{t-1} + (1 - \alpha)S_{t-2}]$$
$$= \; \alpha y_t + \alpha(1 - \alpha)y_{t-1} + (1 - \alpha)^2 S_{t-2},$$

which can be iterated to show that the coefficient of y_{t-k} is $\alpha(1 - \alpha)^k$.

The EWMA may be written also in terms of its preceding value and the deviation of the current observation from the preceding value as

$$\text{EWMA}_t \; = \; \alpha y_t + (1 - \alpha)\text{EWMA}_{t-1}$$
$$= \; \text{EWMA}_{t-1} + \alpha[y_t - \text{EWMA}_{t-1}].$$

8.3.4   Using a Moving Average for Prediction

8.3.4.1   Smoothed Value as a Predictor of the Next Value

Whatever the method of smoothing, the smoothed value S_t can be taken as a prediction $\hat{y}_{t+1}$ of the next future value. That is, one way of predicting is to predict the next value of the series to be the current smoothed value, that is,

$$\hat{y}_{t+1} = S_t.$$

For example, the moving-average predictor for the next observation y_{t+1} based on the four most recent observations is

$$\hat{y}_{t+1} = (y_t + y_{t-1} + y_{t-2} + y_{t-3})/4.$$

8.3.4.2   A Predictor-Corrector Formula

It is noted that such a formula *lags*. We can correct such a predictor to form a more responsive, *predictor-corrector* formula. This includes the predicted value as if it were a data point. For four points, for example, it is

$$
\begin{aligned}
\text{Corrected } \hat{y}_{t+1} &= (\hat{y}_{t+1} + y_t + y_{t-1} + y_{t-2})/4 \\
&= [(y_t + y_{t-1} + y_{t-2} + y_{t-3})/4 + y_t + y_{t-1} + y_{t-2}]/4 \\
&= (y_t + y_{t-1} + y_{t-2} + y_{t-3} + 4y_t + 4y_{t-1} + 4y_{t-2})/16 \\
&= (5y_t + 5y_{t-1} + 5y_{t-2} + y_{t-3})/16.
\end{aligned}
$$

This can be applied to a weighted moving average (WMA) such as $.4\,y_t + .3\,y_{t-1} + .2\,y_{t-2} + .1\,y_{t-3}$, giving

$$
\begin{aligned}
\text{Corrected } \hat{y}_{t+1} &= .4\hat{y}_{t+1} + .3y_t + .2y_{t-1} + .1y_{t-2} \\
&= .4(.4y_t + .3y_{t-1} + .2y_{t-2} + .1y_{t-3}) + .3y_t + .2y_{t-1} + .1y_{t-2}. \\
&= .46y_t + .32y_{t-1} + .18y_{t-2} + .04y_{t-3}.
\end{aligned}
$$

In the exercises, the reader is asked to compute the predictor-corrector formula for five-point and six-point sum-of-digits WMAs.

8.3.4.3   MACD

An application of a difference between a short-term and longer-term moving average is considered next. Such differences are used in a number of fields, including epidemiology and financial investments analysis. *Moving Average Convergence/Divergence* (MACD) —"Mac D"—is an indicator of movement in the price of a security. MACD is the difference, a short-term EMA minus a long-term EMA. Here t is usually in days, that is, one averages daily closing prices. The smoothing constant α is often taken to be of the form $2/(n+1)$, where n is the period of the EMA. An EMA with α equal to $2/(n+1)$ is considered as roughly equivalent to a moving average of n days. If $n = 1$, then $2/(n+1) = 1$, and the EMA is simply y_t. If $n = 2$, then $2/(n+1) = 2/3$; then the EMA is

$$\tfrac{2}{3}y_t + \tfrac{2}{9}y_{t-1} + \tfrac{2}{27}y_{t-2} + \cdots \approx .67y_t + .22y_{t-1} + .07y_{t-2} + \cdots,$$

which is close to the sum-of-digits weighted moving average for $n = 2$, namely

$$(2\,y_t + 1\,y_{t-1})/3 = (2/3)\,y_t + (1/3)\,y_{t-1}.$$

A typical value of α for the short-term EMA in a MACD might be $\alpha = 2/(12{+}1) = 2/13 \approx .15$ and the long-term EMA's α might be $\alpha = 2(/26{+}1) = 2/17 \approx .074$, approximately half of the short-term α. These EMAs are turned into a *momentum oscillator* by subtracting the longer moving average from the shorter moving average. The resulting plot of the difference against t forms a line that oscillates above and below zero.

There are many MACD formulas. Using shorter EMAs (ones with higher α) will produce quicker, more responsive indicators, while using longer moving averages will produce slower indicators, less prone to anomalous patterns such as whipsaws, which occur when a buy or sell signal is reversed in a short time.

Usually, a 9-day EMA of MACD itself is plotted along side to act as a trigger line. A bullish crossover occurs when MACD moves above its 9-day EMA, and a bearish crossover occurs when MACD moves below its 9-day EMA. Note that a month includes about 22 trading days, so 26 trading days is just a little more than a month. Twelve days is about two and a half trading weeks. Nine days is just under two trading weeks. One simple signal that is used is when the current price crosses the nine-day EMA.

There seems to be no particular basis for the choices 12, 26, and 9. Further, MACD does not seem necessarily to be effective in producing excess returns, that is, rates of return above what is generating by a benchmark strategy such as buy-and-hold. Even modest transaction costs can wipe out the returns shown by MACD (St. John 2010).

8.4 Need for Modeling

There are so many ways to form moving averages, depending upon the window width and the weights, that one feels a need to develop a model to describe the process generating the data and then develop the method accordingly. This modeling approach is taken up in subsequent sections. Faced with such a wide choice of moving averages and the like, the discussion now turns to statistical modeling of time series, following from a mathematical description of the process generating the series, in the hope that such modeling will enable the generation of optimal procedures corresponding to the various models.

8.5 Trend, Seasonality, and Randomness

One way of modeling is to consider a time series as containing a *trend* (upward or downward drift), seasonality, and randomness. Economic statisticians sometimes consider *multiplicative models* such as

$$z_t = A_t B_t \delta_t, \ t = 1, 2, \ldots, n,$$

where A_t is seasonal, B_t is trend, and δ_t is multipicative random error. Using the fact that the log of a product is the sum of the logs, write

$$\log z_t = \log(A_t B_t \delta_t) = \log A_t + \log B_t + \log \delta_t$$

or

$$Y_t = \alpha_t + \beta_t + \varepsilon_t,$$

where $Y_t = \log z_t$, $\alpha_t = \log A_t$, $\beta_t = \log B_t$, $\varepsilon_t = \log \delta_t$.

For example, suppose that the data are quarterly. Take

$$\alpha_t = \gamma_1 x_{1t} + \gamma_2 x_{2t} + \gamma_3 x_{3t} + \gamma_4 x_{4t},$$

where, for $j = 1, 2, 3, 4$, the dummy variable $x_{jt} = 1$ for values of t that are in the j-th quarter and 0 for other values of t. The trend might be taken to be linear, $\beta_t = \beta t$. The parameter β represents the average increase per quarter. This gives the model

$$Y_t = \gamma_1 x_{1t} + \gamma_2 x_{2t} + \gamma_3 x_{3t} + \gamma_4 x_{4t} + \beta t + \varepsilon_t, \ t = 1, 2, \ldots, n.$$

For example,

$$
\begin{aligned}
\mathcal{E}[Y_1] &= \beta + \gamma_1, \text{ one quarter of trend plus first-quarter effect} \\
\mathcal{E}[Y_2] &= 2\beta + \gamma_2, \text{ two quarters of trend plus second-quarter effect} \\
\mathcal{E}[Y_3] &= 3\beta + \gamma_3, \text{ three quarters of trend plus third-quarter effect} \\
\mathcal{E}[Y_4] &= 4\beta + \gamma_4, \text{ four quarters of trend plus fourth-quarter effect} \\
\mathcal{E}[Y_5] &= 5\beta + \gamma_1, \text{ five quarters of trend plus first-quarter effect.}
\end{aligned}
$$

Remarks. (i) Such a model can be appropriate if the assumption of uncorrelated errors ε_t is satisfied. This should be examined using the autocorrelation function of the residuals. (See the section on autocorrelation functions later in the chapter.) (ii) Sometimes the trend will be eliminated by differencing. If the mean of Y_t contains a term βt, then the mean of Y_{t-1} contains a term $\beta(t-1)$, and the mean of the difference $Y_t - Y_{t-1}$ will contain a term $\beta t - \beta(t-1) = \beta$, a constant not involving t, so that the mean of the difference can be level. If a series is uptrending, it is tempting to model it using regression on t, or on t and t^2. Although a dataset can be fitted in such a manner, it is illogical to do so, because phenomena (economic, financial, biological) are not unlimited, whereas t goes to infinity, and using t^2 will lead to a parabola that goes up or down to infinity as well; and similarly for higher order polynomials. It is usually preferable to *difference* the data. This means analyzing differences such as $Y_t - Y_{t-1}$ instead of Y_t itself. It is just as natural to consider these differences, or changes, as it is to analyze the process itself. (See also the section on pre-processing the data later in the chapter.) (iii) Seasonal effects are discussed later in the chapter in the section on seasonal data and the section on dynamic regression models.

8.6 Models with Lagged Variables

Next considered is the problem of bringing past values of variables into the model. These may be past values of the same dependent variable Y or past values of other, explanatory variables, or both.

8.6.1 Lagged Variables

Given a time series consisting of observed values $\{y_t, t = 1, 2, \ldots, n\}$, denote the corresponding variable by Y_t. It is helpful to distinguish between a variable and its values. When convenient, the value of the variable upper-case Y is denoted by lower-case y.

The *lagged variable* is denoted by Y_{t-1}. If Y_t is today's value, Y_{t-1} denotes yesterday's value. At time t, $Y_t = y_t$ and the lagged variable $Y_{t-1} = y_{t-1}$. The lag-two variable is denoted by Y_{t-2}. It would be the result from the day before yesterday. The lag-two variable is the lag of the lag-one variable. The lag-k variable is denoted by Y_{t-k}.

A correlation between Y_t and Y_{t-k} is called an *autocorrelation*. The prefix "auto-" means "self," and this is a correlation between Y_t and a lagged version of itself.

8.6.2 Autoregressive Models

The model

$$Y_t = \phi_0 + \phi_1 y_{t-1} + \phi_2 y_{t-2} + \cdots + \phi_p y_{t-p} + \varepsilon_t$$

is an *autoregressive model* (*autoregression model*) of order p. It involves lags up to and including order p. The lagged values are written as lower-case because at time t they are realized (known, observed) values. The *autoregression coefficients* are $\phi_1, \phi_2, \ldots, \phi_p$. The model is analogous to the multiple regression model $Y_i = \beta_0 + \beta_1 x_{1i} + \beta_2 x_2, + \cdots + \beta_p x_{pi} + \varepsilon_i$. The autoregression coefficients are denoted by ϕ instead of β, and the symbol ϕ_0, rather than α or β_0, denotes the constant (intercept) in the model. The errors ε_t are "white noise." A *white noise* sequence is uncorrelated: $\mathcal{C}[\varepsilon_t, \varepsilon_u] = 0$, for all $t, u, t \neq u$, and has common variance (variance not varying with t) $\mathcal{V}[\varepsilon_t] = \sigma_\varepsilon^2$.

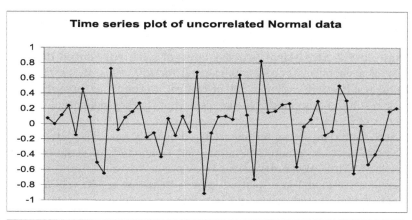

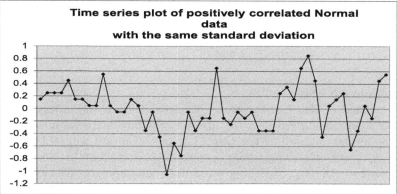

FIGURE 8.1

Uncorrelated and positively correlated data

To compare and contrast correlated and uncorrelated data, Figure 8.1 shows simulated uncorrelated and correlated data with the same mean (just below 5) and standard deviation (just below 0.4). The autocorrelation coefficient is positive. When the correlated process goes up, it tends to go up for a while; when it goes down, it tends to go down for a while. And if one observation is above the center line, chances are that the next one will be too. The plot of the uncorrelated data is jumpier; two successive observations are likely to be on the opposite sides of the mean. The plot of the positively correlated data is smoother than this. (This would not be the case, however, if the autocorrelation coefficient were negative.. Then the plot would be even more jumpy than uncorrelated data, and jagged when successive points are connected with a line.)

Next, some examples of autoregression will be considered; one in quality control, another in financial investments analysis, and a third in economics.

Example 8.2 Statistical process control with correlated data

Yield of a chemical process was charted for fifty runs of the process, and several results were declared out of control, beyond the three-sigma limits. These results were too far from the mean, but do they signal that the process was out of control?

Such limits are set up for uncorrelated observations. The data were found to have a first-order autocorrelation that accounted for some upward and downward drifts. When deviations from predicted values based on a simple autoregressive model were plotted, one of the same results still appeared to be out of control, but others did not. Further, a couple of other points, not detected by the original three-sigma limits, appeared to be out of control. So, there had been both false-alarms and failures to detect that were corrected by a more appropriate model.

Example 8.3 Day trading

In the kind of *day trading* to be discussed here, if the investor trades a given stock on a given day, the investor will buy shares of it at the open of the market and sell at the close. The rate of return will be the *intraday rate of return,* the change from open to close, divided by the opening price, $Q_t = (C_t - O_t)/O_t$. It is interesting to fit an autoregressive model to Q_t, that is, to regress intraday ROR on its lag to see if there might be a hint of a way to do effective day trading. This will be introduced now and discussed further later in the chapter, where day-of-the-week effects will also be taken into account.

An *exchange-traded fund* (ETF) is like a mutual fund, consisting of assets like stocks, commodities, or bonds, that trades on an exchange. ETFs offer diversification like a mutual fund but trade like ordinary stocks. Some ETFs track an index, such as the S&P500. This example concerns MDY, the S&P400 EFT of stocks of 400 midcap companies. (A similar analysis was done in Ortiz (2008), in a dissertation advised by the author.) We considered MDY because we thought it was a bit more exciting than the ETF of SPDR, based on the S&P500, which are higher cap. The data here are from 1995: Aug. 18 through 2007: June 21. This is 2,981 observations, 2,980 RORs. We used continuous RORs, although most of the values are small so that there would be little difference between continuous ROR and ordinary ROR. (The symbol Q_t rather than q_t is used here for continuous ROR because here q_t means the realized value of Q_t.)

Here are some summary statistics.

```
Descriptive Statistics: intradayROR (as decimal, not pct)
-----------------------------------------------------------------

    N          Mean    SE Mean     StDev     Minimum            Q1
-----------------------------------------------------------------

  2980     -0.000409  0.000200  0.010904   -0.087640    -0.006665
---------------------------------------

  Median         Q3   Maximum
---------------------------------------
0.000280   0.005945  0.064320
```

The mean is -0.0409%; its standard error is $s_y/\sqrt{n} = 1.0904\%/\sqrt{2980} \approx 1.0904/54.59 = 0.0200\%$. The five-point summary is

$$
\begin{aligned}
min &= -8.7\%, \\
Q1 &= -0.67\%, \\
median &= +0.028\%, \\
Q3 &= +0.059\%, \\
max &= +6.4\%.
\end{aligned}
$$

The range is max - min $= +6.4\% - (-8.7\%) = 15.1\%$ but the interquartile range is only Q3 - Q1 $= +0.059\% - (-0.67\%) = 0.73\%$.

For now, results from a simple first-order autoregression are shown (although the model identification procedures presented later in the chapter suggest perhaps two terms, either second-order autoregression, second-order moving average, or one autoregressive and one moving average term; somewhat more complex models will be considered later in the chapter). The mean square error of fit is 0.000118875; the standard error of fit is $s_\varepsilon = \sqrt{0.000118875} = .010903$, or about 1%. (This is not much smaller than the overall standard deviation, which is .010904.) The estimated autoregression coefficient has a value of 0.0212, with $t = 1.16$ and $p = .247$. This is not particularly significant. However, there is a difference between statistical significance and practical significance. Usually this is stated in the case where a result is statistically significant but not large enough to be of practical importance. Here, the reverse may be true. Although the effect is small, it may represent something real, which, if exploited repeatedly, may lead to some gain.

```
ARIMA model:  IntradayROR

Estimates of Parameters
Type           Coef    SE Coef     t      p
AR   1       0.0212     0.0183   1.16  0.247
```

```
Constant   -0.0004002   0.0001997   -2.00   0.045
Mean       -0.0004089   0.0002040
```

```
Number of observations:   2981
Residuals:     SS =   0.354129 (backforecasts excluded)
               MS =   0.000118875   DF = 2979
```

Decision risk analysis. The predicted value is $\hat{Q}_{t+1} = -0.04002\% + 0.0212\,q_t$. The investor might execute a day trade tomorrow if the predicted value is sufficiently large. When a buy or a sell is executed, there is a transaction cost TC. If TC is expressed as a rate, for example .001 or 0.1%, it can be subtracted directly from the ROR to get a net ROR. If the transaction cost is a fixed amount, say \$10 per trade, and the investor will buy, say 1,000 shares at \$100 per share, for \$100,000, and sell the 1,000 shares at the end of the day at \$101.00 per share, for \$101000, then the TC rate is TC $= (10 + 10)/(100,000 + 101,000) = 20/201,000 \approx 20/200,000 - 1/10,000 = .0001$ or 0.01%. Figuring a TC for both buying and selling, the mean daily net ROR is estimated as $(0.035 - 2 \times TC + 0.021\% - 2 \times TC + 3\,IR)/5$. For example, if TC $= 0.1\%$, it could completely erase the positive expected gain on days when the buy/sell is executed.

A decision risk analysis for this situation proceeds as follows. The net gain of investing tomorrow is estimated as $\hat{Q}_{t+1} -$ TC, where TC is the transaction cost, expressed as a rate.

The gain of not investing tomorrow is the one-day interest rate, say IR_{t+1}. IR can be considered as an actual interest payment when this is realistic, or, in any case, as an opportunity cost, unrealized interest, if the investor trades on day $t + 1$. The difference in the two is $(\hat{Q}_{t+1} - TC) - IR_{t+1} = \hat{Q}_{t+1} - (TC + IR_{t+1}) = \hat{Q}_{t+1} - c$, where $c = TC + IR_{t+1}$. This is positive if $\hat{Q}_{t+1}$ exceeds this amount c. So a strategy is to do the day trade tomorrow if the predicted intraday ROR exceeds c. Note that $\hat{Q}_{t+1} = a + b\,q_t$, where $a = \hat{\phi}_0$ and $b = \hat{\phi}_1$. The condition $\hat{Q}_{t+1} > c$ is equivalent to $a + b\,q_t > c$, that is, $q_t > (c - a)/b$ if $b > 0$ or $q_t < (c - a)/b$ if $b < 0$.

The strategy can be run on a spreadsheet for past data and the accumulated gain (or loss) computed. Also, a theoretical computation of the expected net gain can be done to see what might be expected and how it depends on the parameters. The expected net gain is

$$\Pr\{\hat{Q}_{t+1} > c\}\,\mathcal{E}[\,Q_{t+1} - TC\,|\,\hat{Q}_{t+1} > c\,] + \Pr\{\hat{Q}_{t+1} \leq c\}\,IR_{t+1},$$

where IR_{t+1} is treated as a constant as it would be known on day t, when the decision is made. For short, write $\hat{Q}_{t+1}$ as $a + b q_t$, where $a = \hat{\phi}_0$ and $b = \hat{\phi}_1$. When the ROR Q_t has a specified distribution, this can in theory be calculated, for example when Q_{t+1} is distributed according to a Normal distribution with mean $\mu = \phi_0/(1 - \phi_1)$ and variance $\sigma^2 = \sigma_\varepsilon^2/(1 - \phi_1^2)$. The computation involves the conditional expectation of a Normal r.v., given that it exceeds a given constant, which can be evaluated using a result from Appendix B on Normal distributions. We do not pursue this calculation further here.

Further thoughts. A more appropriate data analysis results if the first part of the dataset is used as a training set and the rest as a test set. The training set is for estimating the parameters of the model. The test set is for evaluating performance in future samples. Because estimates are optimized for the training set, there will be shrinkage of goodness of fit and prediction in future samples. Thus, looking at the performance of the estimated model in the test set gives a better idea of what to expect in practice.

A way to implement the strategy would be on a *rolling* basis, updating the estimates each day. Once a spreadsheet or program is written, this is not difficult. A rolling window of perhaps 252 days (one year of trading days) or 504 days might be used.

Example 8.4 Price and quantity

Next, consider an economics example of autoregression. Relationships of price and quantity to lags of one another lead to simultaneous equations and then to an autoregression for price. To see this, note that price can be a function of quantity and at the same time quantity a function of price. This leads to a system of equations such as

$$\mathcal{E}[p_t \mid q_t] = c + dq_t$$

$$\mathcal{E}[q_t \mid p_{t-1}] = a + bp_{t-1}.$$

Substituting the second equation into the first gives the autoregression

$$\mathcal{E}[p_t \mid p_{t-1}] = c + d(a + bp_{t-1}),$$

or

$$\mathcal{E}[p_t \mid p_{t-1}] = \phi_0 + \phi_1 p_{t-1},$$

with $\phi_0 = ad + c$, $\phi_1 = bd$. This is a first-order autoregression of price alone.

A *second-order autoregression* takes the form

$$y_t = \phi_0 + \phi_1 y_{t-1} + \phi_2 y_{t-2} + \varepsilon_t$$

where $\{\varepsilon_t\}$ is white noise. An alternative way to write this model is in terms of the preceding value y_{t-1} and the change $y_{t-1} - y_{t-2}$. To do this, subtract and add y_{t-1} to y_{t-2} and write the autoregressive part of the model as

$$
\begin{aligned}
\phi_1 y_{t-1} + \phi_2 y_{t-2} &= \phi_1 y_{t-1} + \phi_2(y_{t-2} - y_{t-1} + y_{t-1}) \\
&= \phi_1 y_{t-1} + \phi_2 y_{t-1} + \phi_2(y_{t-1} - y_{t-2}) \\
&= (\phi_1 + \phi_2)y_{t-1} + \phi_2(y_{t-1} - y_{t-2}).
\end{aligned}
$$

The difference $y_{t-1} - y_{t-2}$ occurring between times $t-2$ and $t-1$ can change signs, producing ups and downs in the pattern of the series.

8.7   Moving-Average Models

How would you model a time series where adjacent observations are correlated, but observations more than one time period apart are uncorrelated? Try a model $Y_t = \mu + a_t$, where the errors a_t are written in terms of a white noise sequence $\{\varepsilon_t\}$. In this simplest case, the error sequence is $a_t = \varepsilon_t - \theta\varepsilon_{t-1}$. This gives

$$
\begin{aligned}
\mathcal{C}[Y_t, Y_{t-1}] &= \mathcal{C}[\mu + a_t, \mu + a_{t-1}] = \mathcal{C}[a_t, a_{t-1}] \\
&= \mathcal{C}[\varepsilon_t - \theta\varepsilon_{t-1}, \varepsilon_{t-1} - \theta\varepsilon_{t-2}] \\
&= \mathcal{C}[\varepsilon_t, \varepsilon_{t-1}] + \mathcal{C}[\varepsilon_t, -\theta\varepsilon_{t-2}] \\
&\quad + \mathcal{C}[-\theta\varepsilon_{t-1}, \varepsilon_{t-1}] + \mathcal{C}[-\theta\varepsilon_{t-1}, -\theta\varepsilon_{t-2}] \\
&= 0 + 0 - \theta\,\mathcal{C}[\varepsilon_{t-1}, \varepsilon_{t-1}] + 0 \\
&= -\theta\,\mathcal{C}[\varepsilon_{t-1}, \varepsilon_{t-1}] = -\theta\,\mathcal{V}[\varepsilon_{t-1}] = -\theta\,\sigma^2.
\end{aligned}
$$

It is easy to see that $\mathcal{V}[Y_t] = (1 + \theta^2)\sigma^2$ and hence that $\text{Corr}[Y_t, Y_{t-1}] = -\theta/(1 + \theta^2)$. For $k = 2, 3, \ldots, \mathcal{C}[Y_t, Y_{t-k}] = 0$.

A second-order MA model would have $a_t = \varepsilon_t - \theta_1\varepsilon_{t-1} - \theta_2\varepsilon_{t-2}$. An MA model of order q ($q = 1, 2, 3, \ldots$) is analogously defined. Then observations one or two time periods apart would be correlated, but observations three or more time periods apart would be uncorrelated. The reader can consult one or another of the texts in the Bibliograpy for the development of higher-order MA and AR models.

8.7.1   Integrated Moving-Average Model

Consider the particular moving-average model,

$$ Y_t = y_{t-1} + a_t, $$

where the errors a_t are correlated and described as $a_t = \varepsilon_t - \theta\varepsilon_{t-1}$, the sequence $\{\varepsilon_t\}$ being "white noise"—uncorrelated and with equal variance σ_ε^2. These white-noise variables ε_t are like those used in ordinary simple and multiple regression models. Often they are taken to be Normal, in which case we refer to "Gaussian white noise."

This particular moving-average model is a reasonable model for many phenomena. It is a "random walk," in that the current value is equal to the preceding value plus random error. Here we allow the error sequence to be correlated.

Another way to consider this model is that the difference is equal to correlated random error. That is, letting $w_t = Dy_t = y_t - y_{t-1}$, the model is $w_t = \varepsilon_t - \theta\varepsilon_{t-1}$. Such a model is called an *integrated model* because the difference is directly modeled in terms of the basic building blocks, the error sequence; the observed series is an "integrated" version of this. In the

Box/Jenkins notation IMA(d, q), this model is denoted by IMA(1,1), meaning that q, the order of MA, is 1, and d, the order of differencing, is 1. (See below for a more complete explanation of this notation.)

This integrated moving-average model leads to exponential smoothing as a prediction procedure. To see this, note that the model for Y_{t+1} is

$$Y_{t+1} = y_t + \varepsilon_{t+1} - \theta\dot{\varepsilon}_t.$$

This gives

$$\hat{y}_{t+1} = y_t - \hat{\varepsilon}_{t+1} - \theta\hat{\varepsilon}_t = y_t + 0 - \theta(y_t - \hat{y}_t),$$

because the predicted value of the next error is 0 and the predicted value of the current error is the difference between the current value and its prediction. This gives

$$\hat{y}_{t+1} = y_t - \theta(y_t - \hat{y}_t) = (1 - \theta)y_t + \theta\hat{y}_t = \alpha y_t + (1 - \alpha)\hat{y}_t,$$

where α is the smoothing constant discussed above. That is to say, the EWMA smoothing constant can be estimated by fitting this model and taking α to be one minus the estimate of θ.

8.7.2 Preliminary Estimate of θ

Let $W_t = Y_t - Y_{t-1}$. As shown above, the first-order autocorrelation ρ_1 of W_t is $-\theta/(1 + \theta^2)$. An estimate of θ can be obtained by the method of moments, setting ρ_1 equal to its sample analog, r_1, the sample lag-one autocorrelation of W_t :

$$-\hat{\theta}/(1 + \hat{\theta}^2) = \hat{\rho} = r_1.$$

This gives a quadratic equation, the two roots of which are

$$\hat{\theta} = -1/(2r_1) \pm [1/(2r_1)^2 - 1]^{1/2}.$$

The two solutions are reciprocals, so only one will be less than one in size, the other solution being extraneous. If, for example, $r_1 = -12/25 = -.48$, these solutions are $4/3$ and $3/4$, so $\hat{\theta}$ is $3/4$, and the estimate of the smoothing constant α is $1 - \hat{\theta} = 1/4$. This could now be used, for example, in Excel's exponential smoothing command. (Note, however, that what Excel calls the "smoothing constant" is not our α but rather our $\theta = 1 - \alpha$.)

8.7.3 Estimate of θ

This method of estimation by quadratic equation does not work for all values of r_1; statistical computer programs use methods that work for a wider range of values of r_1. So, put your data into such a program, go to the procedures for Box/Jenkins ARIMA models, and enter the order p of autoregression as 0, the order d of differencing as 1, and the order q of the moving average part as 1. Get the estimate of θ, and then subtract it from 1 to get the estimate of the smoothing constant α. But, once you are in the statistical computer program, you can get it to do the forecasting that you want, so you would not necessarily need to take the value of α back to a spreadsheet.

8.7.4 Integrated Moving-Average with a Constant

Next, suppose the model is

$$Y_t = y_{t-1} + \delta + \varepsilon_t - \theta\varepsilon_{t-1}.$$

Now a constant δ has been included in the model. In terms of the difference variable W_t, this is

$$W_t = \delta + \varepsilon_t - \theta\varepsilon_{t-1}.$$

The parameter δ is the mean of W_t: $\delta = \mu_w$. The time series W_t is supposed to be level, but its mean δ may be non-zero. This implies that the series Y_t has a drift of an amount δ per time period.

To deal with this, suppose that Z_t is zero-mean MA(1). This variable can be obtained by subtracting the mean δ, estimated by $\bar{w}$, from each value of the difference w_t. That means that the model for Z_t is

$$Z_{t+1} = \varepsilon_{t+1} - \theta\dot{\varepsilon}_t.$$

This gives

$$\hat{z}_{t+1} = -\hat{\varepsilon}_{t+1} - \theta\hat{\varepsilon}_t = 0 - \theta(z_t - \hat{z}_t) = -\theta(z_t - \hat{z}_t),$$

because the predicted value of the next error is 0 and the predicted value of the current error is the difference between the current value and its prediction.

So the model with constant can be handled by

- Getting $\hat{z}_{t+1}$ as $\hat{z}_{t+1} = (1 - \theta)z_t + \theta\hat{z}_t$,

- Adding $\bar{w}$ to get $\hat{w}_{t+1} = \hat{a}_{t+1} + \bar{w}$,

- Getting $\hat{y}_{t+1} = \hat{w}_{t+1} + y_t$.

8.8 Identification of ARIMA Models

A wide class of models for time series is the class of Box/Jenkins ARIMA models (Box and Jenkins 1970, 1976; Box, Jenkins, and Reinsel 1994). ARIMA means autoregressive, integrated, moving average, AR for autoregression, I for integrated, MA for moving average.

8.8.1 Pre-Processing

Pre-processing may involve transformation or differencing.

8.8.1.1 Transformation

Sometimes as a first step in the analysis, the series is transformed; for example, the logarithm $\ln Y$ instead of Y may be analyzed. Such is often the case when Y varies over several orders of magnitude.

8.8.1.2 Differencing

As mentioned above, it may be necessary to difference the series before computing the ACF and PACF. This is because the mathematical theory of stochastic processes states that a *stationary* process may be represented as an AR or an MA. Part of the idea of stationarity is that the distribution of Y_t should not depend upon t. In particular, the series must be level, that is, not uptrending or downtrending. So a first step in analyzing a series is to examine a plot of it to see if it is trending. If so, the series is differenced and the differences are analyzed. Often, the first difference $\mathcal{D}[Y_t] = Y_t - Y_{t-1}$ is sufficient.

Example 8.5 Continuous rate of return

Let P_t be the price of a share of stock at time t. Often in the time period under consideration the price may have a wide range. Then the log transform $L_t = \ln P_t$ may be made. If the series is uptrending, then the difference may be computed, resulting in $\mathcal{D}L_t = L_t - L_{t-1} = \ln P_t - \ln P_{t-1}$. This is the continuous ROR.

Sometimes a *second* difference is used. This would be the case if the first difference is still trending. The second difference $\mathcal{D}^2 Y_t$ is the difference of the difference.

$$
\begin{aligned}
\mathcal{D}^2 Y_t &= \mathcal{D}(\mathcal{D}Y_t) = \mathcal{D}W_t = W_t - W_{t-1} \\
&= (Y_t - Y_{t-1}) - (Y_{t-1} - Y_{t-2}) \\
&= Y_t - 2\,Y_{t-1} + Y_{t-2}.
\end{aligned}
$$

Example 8.6 Velocity and acceleration

If Y_t is the position of an object moving along the axis at time t, then $\mathcal{D}Y_t$ is the velocity from time $t-1$ to time t, and

$$
\mathcal{D}^2 Y_t = \mathcal{D}(\mathcal{D}Y_t) = (Y_t - Y_{t-1}) - (Y_{t-1} - Y_{t-2}) = Y_t - 2\,Y_{t-1} + Y_{t-2}
$$

is the change in velocity, or the acceleration, from time $t - 2$ to time t.

8.8.2   ARIMA Parameters p, d, q

The parameter d denotes the order of differencing, p the order of autoregression, and q the order of the moving-average part of the model, if any. These are integers, $d = 0, 1, 2, \ldots$, $p = 0, 1, 2, \ldots$, $q = 0, 1, 2, \ldots$. Usually these are each 0, 1, or 2. First, the order d of differencing required to obtain a level series is determined. Then the autocorrelation function ACF and partial autocorrelation PACF are used to choose p and q. (ACF and PACF are discussed in the next section.) Then the values of p, d, q are supplied to the software, and estimation and forecasting proceed. SCA software (Liu 2008) has an expert system command that will automatically identify the model, that is, automatically choose p, d, q.

A shorthand notation for the model is used when one or another of p, d, q is zero.

- ARIMA$(p, 0, q)$ = ARMA(p, q)
- ARIMA$(0, d, q)$ = IMA(d, q)
- ARIMA$(p, d, 0)$ = ARI(p, d)
- ARIMA$(0, 0, q)$ = MA$(q$)
- ARIMA$(p, 0, 0)$ = AR(p)

For example, the integrated moving average model discussed above is denoted by IMA(1,1), that is, $d = 1$, and $q = 1$.

- A in ARIMA is for Autoregression, discussed above.

- I in ARIMA is for Integration, referring to the fact that the observed series may be an integration of a level series and may have to be differenced before analysis.

- MA in ARIMA is for Moving Average, referring to the part of the model capturing the autocorrelation of the errors.

8.8.3   Autocorrelation   Function;   Partial   Autocorrelation Function

Statistical software has procedures for identifiying and estimating Box/Jenkins ARIMA models. First, the given time series must be differenced to obtain a level series. Then the autocorrelation function ACF and the partial autocorrelation function PACF of the differenced series are computed.

The ACF (*autocorrelation function*) is the set of autocorrelations ACF(k) for a number of lags $k = 1, 2$, etc. ACF(k) is the ordinary correlation between an X and a Y where Y is Y_t and X is Y_{t-k}.

The PACF (*partial autocorrelation function*) is the set of partial autocorrelations PACF(k) for a number of lags $k = 1, 2$, etc. PACF(k) is ACF(k), with the effects of intervening lags 1, 2, $k-1$ removed, that is, it is the partial correlation between Y_t and Y_{t-k}, adjusting for $Y_{t-1}, Y_{t-2}, \ldots, Y_{t-k+1}$.

- If the process is MA(q), the ACF has spikes at lags 1 through q and the PACF tails off.

- If the series is AR(p), the ACF tails off and the PACF has spikes at lags 1 through p.

- If the series is ARMA(p, q), the ACF has an irregular pattern with some high and some low values at lags 1 through q, then tails off, and the PACF tails off.

In software, the user can specify a max number of lags for the computation, or use the default. In Minitab, for example, the default is $n/4$ for a series with $n \leq 240$ or $\sqrt{n} + 45$ for $n > 240$.

Charts indicating the method for time-series model identification are shown: Table 8.1 is for MA; Table 8.2, for AR; and Table 8.3, for ARMA. These are oriented horizontally; such charts can be oriented vertically as well. The number of lags shown is seven. The figures suggest three spikes, or order 3; in practice, 1 or 2 would be more typical.

TABLE 8.1
ACF and PACF Pattern for MA(q)

lag:	ACF	PACF
1	X X X X X X X	X X X X X X X
2	X X X X X X	X X X X X
3	X X X X X X	X X X X
4	X	X X X
5	X	X X
6	X	X
7	X	X
	Spikes at lags 1 to q, then cuts off.	Tails off.

8.9   Seasonal Data

Seasonal data data include *monthly* and *quarterly* data. For financial accounting, the *quarters* of the year are three-month periods ending March 31, June 30, September 30, and December 31. It is natural to use the term "seasonal

TABLE 8.2
ACF and PACF Pattern for AR(p)

lag:	ACF	PACF
1	X X X X X X X	X X X X X X X
2	X X X X X	X X X X X X
3	X X X X	X X X X X X X
4	X X X X	X
5	X X X	X
6	X X	X
7	X	X
	Tails off.	Spikes at lags 1 to p, then cuts off.

TABLE 8.3
ACF and PACF Pattern for ARMA(p, q)

lag:	ACF	PACF
1	X X X X X X X	X X X X X X X
2	X X X X X X	X X X X X
3	X X X X X X X	X X X X
4	X X X	X X X
5	X X	X X
6	X X	X
7	X	X
	Irregular at lags 1 to q, then tails off.	Tails off.

data" for such data, with the quarters more or less corresponding to Winter, Spring, Summer, and Fall.

The idea for statistical description and modeling is that

- This year's Fall quarter sales may be related to last year's Fall quarter sales

- This year's Winter sales, to last year's Winter sales, etc.

The designation *seasonal* refers also to time periods other than the seasons. For example, monthly data are seasonal, as are weekly data. Data observed daily, by day of the work week (M, T, W, R, F), are seasonal, as are data observed on all seven days (M, T, W, R, F, S, N).

8.9.1   Seasonal ARIMA Models

The modeling may be done by treating the seasons as categories or by using regression (seasonal autoregression). In the case of quarterly data, one might regress Y_t on Y_{t-4} or Y_{t-4} and Y_{t-8}. After differencing an increasing series Y_t by computing the quarterly difference $W_t = Y_t - Y_{t-4}$, an ARMA model may be fit to the differences W_t.

Example 8.7 Best Buy company quarterly sales

For many retail firms, fourth-quarter sales are highest. For Best Buy, they typically account for about a third of annual sales. Thus there are strong quarterly effects.

Best Buy revenue has been uptrending, with an average annual increase of about 20%. (More recently, this growth has diminished.) When a series is not level, then as time goes on, more and more parameters may be required to model the series adequately. In such a case, the differences, that is, the changes from one time to the next, can be analyzed. Here these are the quarterly differences, the increases from last year's q-th quarter to this year's q-th quarter, $q = 1, 2, 3, 4$. The quarterly difference $\mathcal{D}4_t = Y_t - Y_{t-4}$ becomes the response variable in a regression analysis. The explanatory variable is $\mathcal{D}4_{t-4}$, which is $Y_{t-4} - Y_{t-8}$. That is, the quarterly difference $\mathcal{D}4_t$ is analyzed in a first-order seasonal autoregressive model, $\mathcal{D}4_t = \Phi_0 + \Phi_1 \mathcal{D}4_{t-4} + \varepsilon_t$.

Update. In early 2009, Best Buy's sales increase was aided by the closing of a competitor, Circuit City. Best Buy quarterly sales are further analyzed in the section below on *stable seasonal pattern* (Section 8.9.2).

Seasonal ARIMA parameters. If the data are quarterly, the value of Y_t could be expected to be similar to that of Y_{t-4}. The *seasonality* S is equal to 4.

Monthly data are also seasonal. The seasonality S is 12. The value of Y_t could be expected to be similar to that of Y_{t-12}.

Daily data, such as closing stock prices, on Monday, Tuesday, Wednesday, Thursday, and Friday, are also called *seasonal*. The seasonality S equals 5. For a variable observed every day, Monday through Sunday, the seasonality $S = 7$.

The order of seasonal AR is denoted by upper-case P; that of seasonal MA, by upper-case Q. The overall notation for an ARIMA model with regular and seasonal parts is   ARIMA(p, d, q) S ARIMA(P, D, Q). Similarly, seasonal AR and MA parameters are denoted by upper-case Φ and Θ, respectively.

Example 8.8 U.S. Gross Domestic Product

Here the variable Y_t is quarterly GDP. GDP is reported in billions (10^9) of dollars. So a value 1,000 is one trillion (10^{12}). Here we report our analysis of data from 1947:1 through 2004:4 (58 years, that is, $n = 232$ quarters). Over these six decades, quarterly GDP ranged from about 56 to 3,060 (annual from about 200 to about 12,000). Because the variable ranges over several orders of magnitude, it makes sense to take logs, using the transform $W_t = \ln Y_t$. Also, because the variable is uptrending, it makes sense to difference (quarterly). So the variable

$$W_t = Z_t - Z_{t-4} = \ln Y_t - \ln Y_{t-4} = \ln \mathrm{GDP}_t - \ln \mathrm{GDP}_{t-4}$$

is analyzed. Note that this difference of logs is just the continuous growth rate over four quarters. The fitted model involves this transformation and differencing, and then fitting AR and MA terms. The ACF and PACF of W_t are examined. They both show spikes for lags 1, 2, and 3 and do not cut off immediately after that. So both AR and MA terms are considered. Following Liu (2008, pages 4–13), an ARMA model is fit to the quarterly difference W_t with $Q = 1$ (seasonal first-order MA) and $p = 3$ (regular third-order AR). For $Z_t = \ln \mathrm{GDP}_t$, this is

$$\mathrm{ARIMA}(p = 3, d = 0, q = 0) \quad S = 4 \quad \mathrm{ARIMA}(P = 0, D = 1, Q = 1).$$

- The *seasonality* is $S = 4$

- The *regular part* is $\mathrm{ARIMA}(p = 3, d = 0, q = 0)$,

- The *seasonal part* is $\mathrm{ARIMA}(P = 0, D = 1, Q = 1)$.

```
ARIMA Model: LN GDP
Estimates of Parameters
Type            Coef     SE Coef      t       p
AR   1        1.1163     0.0657   16.98   0.000
AR   2       -0.0156     0.1004   -0.16   0.876 N.S.
AR   3       -0.1715     0.0657   -2.61   0.010
SMA  4        0.5722     0.0636    8.99   0.000
Constant  0.0048221  0.0003980   12.12   0.000

Differencing: 0 regular, 1 seasonal of order 4
Number of observations:
            Original series 232, after differencing 228
Residuals:    SS =  0.0433254 (backforecasts excluded)
              MS =  0.0001943  DF = 223
```

The estimated model is

$$\hat{Z}_t = 0.0048 + 1.1163\,Z_{t-1} - 0.0156\,Z_{t-2} - 0.1715\,Z_{t-3} + e_t - 0.5722e_{t-4}.$$

(Here the symbol e represents the realized value of the error ε. There is a question of how to use the fitted model whose expression includes e terms for prediction. The e terms can also be found in terms of preceding values of the series. For details we defer to books on time series analysis *per se*.) The standard error of fit is $\sqrt{0.0001943} = 0.01394$. This is comparable to Liu's

$$\hat{Z}_t = 0.0609 + 1.044\,Z_{t-1} + 0.018\,Z_{t-2} - 0.338\,Z_{t-3} + e_t - 0.421\,e_{t-4},$$

with a standard error of fit of 0.01737, for the period 1947:1 through 1969:4. (There is a difference of sign in the coefficient of Z_{t-2}, but this coefficient is not s.d.f.z.)

8.9.2   Stable Seasonal Pattern

One way to treat seasonal data is as *compositional data*—considering the whole as the sum of its parts. For example, yearly sales may be considered as composed of (that is, the sum of) monthly sales or quarterly sales. One question to ask is whether the composition (percentages across quarters) is relatively constant ("stable") from year to year. If so, the pattern is said to be a *stable seasonal pattern*. Then forecasting is simplified, in that one can build a model for just the annual totals, forecast the annual total, and apply the appropriate percentages to obtain forecasts for each quarter. The seasonal modeling is separated from the annual modeling (Chen and Fomby 1999).

Example 8.9 Best Buy quarterly sales, continued: stable seasonal pattern

Table 8.4 shows quarterly sales in millions of dollars (M\$) for the eleven years 1998 through 2008, and for the first three quarters of 2009.

Table 8.5 shows the row percents, which are the seasonal pattern. The percentages in the last row (All) are the quarterly totals across years, divided by the grand total. Given a prediction of total sales in 2009, one would predict that 19.71% of those sales would occur in the first quarter. The prediction of total sales in 2009 could be obtained by fitting a time-series model to the annual totals for the years up through 2008. This procedure separates the fitting of a time-series model from the fitting of a seasonal pattern. For example, suppose we use an average annual growth rate to make a forecast for the year 2009. The growth factor has been $45{,}015/8{,}357 = 5.387$, over

TABLE 8.4

Sales, by Quarter (M$)

Year	Q1	Q2	Q3	Q4	Total
1998	1606	1793	2106	2852	8357
1999	1943	2182	2493	3458	10076
2000	2385	2686	3107	4314	12492
2001	2963	3169	3732	5461	15325
2002	3697	4164	4756	6980	19597
2003	4202	4624	5131	6989	20946
2004	5345	5778	6845	7896	25864
2005	6118	6702	7335	10693	30848
2006	6959	7603	8473	12899	35934
2007	7927	8750	9928	13418	40023
2008	8990	9801	11500	14724	45015
Total	52135	57252	65406	89684	264477
2009	10095	11022	12024		

Source. Hoover's; finance.yahoo.com

ten annual increases. This gives an average annual growth factor (multiplier) equal to the tenth root of 5.387, or $5.387^{1/10} = 1.1834$, that is, an average annual factor of .1834 (18.34 % per year). Based on this, at the end of 2008, the forecast of total sales for 2009 would be $45,015 \times 1.1834 = 53,271$ M$.

- The forecast for Q1 of 2009 would have been $19.71\% \times 53,271 = 10,500$ M$, compared to the actual 10,095 M$.

- The forecast for Q2 of 2009 would have been $21.65\% \times 53,271 = 11,533$ M$, compared to the actual 11,022 M$.

- The forecast for Q3 of 2009 would have been $24.73\% \times 53,271 = 13,174$ M$, compared to the actual 12,024 M$.

Later, given the sales of 10,095, 11,022, and 12,024, respectively, for Q1, Q2, and Q3 of 2009, one might forecast the sales for Q4 of 2009 to be

$$53,271 - (10,095 + 11,022 + 12,024) = 20,130 \text{ M\$}$$

Another forecast can be made using the quarterly percentages. The results

TABLE 8.5

Seasonal Pattern: Distribution (%) over Quarters for Each Year

Year	Q1	Q2	Q3	Q4	Total
1998	19.22	21.46	25.20	34.13	100.00
1999	19.28	21.66	24.74	34.32	100.00
2000	19.09	21.50	24.87	34.53	100.00
2001	19.33	20.68	24.35	35.63	100.00
2002	18.87	21.25	24.27	35.62	100.00
2003	20.06	22.08	24.50	33.37	100.00
2004	20.67	22.34	26.47	30.53	100.00
2005	19.83	21.73	23.78	34.66	100.00
2006	19.37	21.16	23.58	35.90	100.00
2007	19.71	21.65	24.73	33.91	100.00
2008	19.97	21.77	25.55	32.71	100.00
All	19.71	21.65	24.73	33.91	100.00

for Q1, Q2, and Q3 total $10{,}095 + 11{,}002 + 12{,}024 = 33{,}121$ M\$. This should be about $19.71 + 21.65 + 24.73 = 66.09\%$ of the total. So another estimate of the total is $33{,}121/.6{,}609 = 50{,}115$ M\$. Then the forecast for Q4 of 2009 would be

$$50{,}115 - 33{,}121 = 16{,}994 M\$.$$

A forecast can be made that combines these via an average or weighted average. The mean of the two forecasts for Q4 of 2009 is $(20{,}130 + 16{,}994)/2 = 18{,}562$ M\$. The weights would be more appropriately chosen proportional to the reciprocal variances (see Chen and Fomby (1999) for details).

There was a recession in the year 2009, so the forecasts may be too high. Another factor is that, according to the theory of life cycle of firms, as a company matures, its growth rate might diminish. Later we continue this case and consider other models for the total annual sales, and other methods for dealing with the quarterly data. But the method of stable seasonal pattern should not be ignored. It is relatively straightforward and simplifies what can be some relatively tricky modeling and forecasting problems.

8.10   Dynamic Regression Models

A regression of Y_t on contemporaneous or lagged values of some explanatory variables X_{1t}, X_{2t}, etc., is regression in a time-series context and thus is called a *dynamic regression model* (or *transfer function model*).

Example 8.10 Steel, iron, and coal

Let s_t, i_t, and c_t be, respectively, the prices of steel, iron and coal in the t-th time period. Iron and coal are used in the production of steel, so one might consider a model such as

$$s_t = \beta_0 + \beta_1 i_{t-1} + \beta_c c_{t-1} + \varepsilon_t,$$

where ε_t is random error.

One could consider a model with a lag of steel price as well as coal and iron prices:

$$s_t = \beta_0 + \beta_s s_{t-1} + \beta_1 i_{t-1} + \beta_c c_{t-1} + a_t.$$

Such models are more complicated than ordinary multiple regression models because the dependent variable appears in lagged form on the right-hand side. However, they may be treated as multiple regression models fairly satisfactorily. (That is, the lagged variable is then treated just as another explanatory variable in the regression.)

Interesting questions arise, including the following.

· If s_{t-1} is in the model, are i_{t-1} and c_{t-1} needed?

· Conversely, if i_{t-1} and c_{t-1} are in the model, is s_{t-1} needed?

Example 8.11 Day trading, continued

As discussed above, a model for intraday ROR can incorporate its lag. Along with this, there could explanatory variables such as ones for day-of-the-week. First, day-of-the-week effects alone can be assessed with an Analysis of Variance (ANOVA). The output is from Minitab, slightly edited. The day-of-week effects are statistically significant, with $F = 2.87$ ($p = .022$). The standard error of fit is $s = 0.01090$. The effects of W and R are positive at $+0.035\%$ and $+0.021\%$; those of M, T, F, negative, at $-.133\%$, -0.112%, and -0.051%. One trading strategy would be to be in on W and R and out on M, T and F.

The mean daily ROR would be $(0.035+0.021)+3\,\mathrm{IR})/5$, where IR is the daily interest rate, assuming that if you let the money sit there, you get interest. For example, if the annual interest rate is 2.6% and the number of business days per year is 260, then IR $= 0.01\%$ per day. The mean daily ROR of the strategy is $(0.035 + 0.021) + 3 \times 0.01)/5 = 0.086/5 = 0.0172\%$ per day.

When a buy or a sell is executed, there is a transaction cost TC. If TC is expressed as a rate, for example .001 or 0.1%, it can be subtracted directly from the ROR to get a net ROR. If the transaction cost is \$ 10 per trade, and the investor will buy, say 1,000 shares at \$100 per share, for \$100,000, and sell the 1,000 shares at the end of the day at \$101.00 per share, for \$101,000, then the rate is $\text{TC} = (10+10)/(100,000+101,000) = 20/201,000 \approx 20/200,000 - 1/10,000 = .0001$ or 0.01%. Figuring a TC for both buying and selling, the mean daily net ROR is estimated as $(0.035 - 2 \times \text{TC} + 0.021\% - 2 \times \text{TC} + 3\,\text{IR})/5$. For example, if TC $= 0.1\%$, it would completely erase the positive expected gain on days when the buy/sell is executed. Suppose then that TC $= 0.005\%$. Then the expected daily net gain is, using IR $= 0.01\%$ per day, $[(0.035 - 2 \times 0.005) + (0.021 - 2 \times 0.005) + 3 \times 0.01)/5 = [0.025 + 0.011 + 3 \times 0.01)/5 = 0.066/5 = 0.0132\%$ per day. There are usually 252 trading days per year. Multiplying by 252 gives $252 \times 0.0132 = 3.3264\%$ per year, a relatively modest amount.

A better test would be obtained by using part of the data as a training set and the more recent part as a test set. Least squares estimates are optimized for the training set, so there will be shrinkage in the goodness of fit when they are applied to the test set.

```
One-way ANOVA: intradayROR versus WEEKDAY

Source      DF        SS        MS     F      p
WEEKDAY      4   0.001363  0.000341  2.87  0.022
Error     2976   0.353853  0.000119
Total     2980   0.355216


s = 0.01090    R-Sq = 0.38%    R-Sq(adj) = 0.25%

                               Individual 95% CIs For Mean Based on
                               Pooled StDev
Level   N      Mean    StDev   --+---------+---------+---------+-------
1      561  -0.00133  0.01179  (--------*--------)
2      612  -0.00112  0.01092     (--------*-------)
3      612   0.00035  0.01069                  (-------*--------)
4      599   0.00021  0.01062                (--------*--------)
5      597  -0.00051  0.01050        (--------*--------)
                               --+---------+---------+---------+-------
                            -0.0020   -0.0010    0.0000    0.0010
Pooled StDev = 0.01090


Tukey 90% Simultaneous Confidence Intervals
All Pairwise Comparisons among Levels of WEEKDAY

Individual confidence level = 98.61%
```

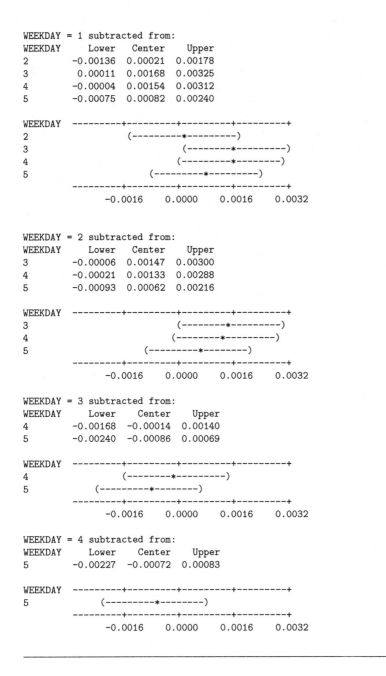

```
WEEKDAY = 1 subtracted from:
WEEKDAY     Lower    Center     Upper
2         -0.00136  0.00021   0.00178
3          0.00011  0.00168   0.00325
4         -0.00004  0.00154   0.00312
5         -0.00075  0.00082   0.00240


WEEKDAY   ---------+---------+---------+---------+
2                  (---------*---------)
3                            (--------*---------)
4                       (---------*--------)
5                   (---------*---------)
          ---------+---------+---------+---------+
              -0.0016    0.0000     0.0016     0.0032


WEEKDAY = 2 subtracted from:
WEEKDAY     Lower    Center     Upper
3         -0.00006  0.00147   0.00300
4         -0.00021  0.00133   0.00288
5         -0.00093  0.00062   0.00216


WEEKDAY   ---------+---------+---------+---------+
3                            (--------*---------)
4                            (--------*---------)
5                       (---------*--------)
          ---------+---------+---------+---------+
              -0.0016    0.0000     0.0016     0.0032


WEEKDAY = 3 subtracted from:
WEEKDAY     Lower    Center     Upper
4         -0.00168  -0.00014   0.00140
5         -0.00240  -0.00086   0.00069


WEEKDAY   ---------+---------+---------+---------+
4                       (--------*---------)
5                   (---------*--------)
          ---------+---------+---------+---------+
              -0.0016    0.0000     0.0016     0.0032


WEEKDAY = 4 subtracted from:
WEEKDAY     Lower    Center     Upper
5         -0.00227  -0.00072   0.00083


WEEKDAY   ---------+---------+---------+---------+
5                   (---------*--------)
          ---------+---------+---------+---------+
              -0.0016    0.0000     0.0016     0.0032
```

Let us now combine day-of-the-week effects with autoregression. An AR model has $\hat{Q}_{t+1} = \hat{\phi}_0 + \hat{\phi}_1 Q_t$. An ANOVA (Analysis of Variance) model with day-of-week can be combined with this in an Analysis of Covariance (ANCOVA).

In multiple regression terms, the model is

$$Q_t = \phi_0 + \phi_1 Q_{t-1} + \alpha_M x_{Mt} + \alpha_T x_{Tt} + \alpha_W x_{Wt} + \alpha_R x_{Rt} + \alpha_F x_{Ft} + \varepsilon_t,$$

$t = 1, 2, \ldots, n$ days, where, for $d = $ M, T, W, R, F, the dummy variable

$$x_{dt} = 1 \text{ if day } t = d \text{ and } = 0 \text{ otherwise.}$$

The analysis can be done on a spreadsheet with a multiple regression function, but then one of the day-of-week terms must be omitted, and ϕ_0 will represent its effect. In statistical software, ANCOVA can be performed with a GLM (General Linear Model) command. In the data spreadsheet, the data are stacked, with a column for the values of the dependent variable Q_t, a column for the lag Q_{t-1}, and a column for the day of the week. The day of the week is entered as the "factor," and the lag is entered as a "covariate." Software output is shown below. One interesting question is whether the correlation of intraday ROR is positive, or negative, from day to day. The estimate of the lag-one autoregression coefficient is positive, at $+0.02113$. The p-value is about .25, not particularly significant. The standard error of fit is $s == 0.0109052$, about the same as that of the ANOVA model.

```
General Linear Model: intradayROR versus WEEKDAY

Factor    Type   Levels  Values
WEEKDAY   fixed       5  1, 2, 3, 4, 5

Analysis of Variance for intradayROR, using Adjusted SS for Tests
Source           DF      Seq SS      Adj SS      Adj MS      F      p
--------------------------------------------------------------------
lagIntradayROR    1   0.0001691   0.0001580   0.0001580   1.33  0.249
WEEKDAY           4   0.0013519   0.0013519   0.0003380   2.84  0.023
Error          2974   0.3536792   0.3536792   0.0001189
--------------------------------
Total          2979   0.3552002

s = 0.0109052   R-Sq = 0.43%   R-Sq(adj) = 0.26%


Term              Coef    SE Coef       t      p
Constant     -0.000469   0.000200   -2.34  0.019
lagIntradayR   0.02113    0.01833    1.15  0.249


WEEKDAY
1            -0.000853   0.000409   -2.09  0.037
2            -0.000625   0.000396   -1.58  0.114
3             0.000841   0.000396    2.12  0.034
4             0.000673   0.000399    1.69  0.092


Means for Covariates
Covariate            Mean     StDev
lagIntradayROR   -0.000472   0.01092


Least Squares Means for intradayROR
```

```
WEEKDAY        Mean   SE Mean
1         -0.001332  0.000460
2         -0.001104  0.000441
3          0.000362  0.000441
4          0.000194  0.000446
5         -0.000514  0.000447
```

Bonferroni 95.0% Simultaneous Confidence Intervals
Response Variable intradayROR
All Pairwise Comparisons among Levels of WEEKDAY

```
WEEKDAY = 1  subtracted from:
WEEKDAY     Lower    Center     Upper  ---+---------+---------+---------+---
2        -0.001563  0.000228  0.002020     (--------*--------)
3        -0.000097  0.001694  0.003485              (-------*--------)
4        -0.000274  0.001527  0.003327             (-------*--------)
5        -0.000984  0.000818  0.002620          (--------*--------)
                                        ---+---------+---------+---------+---
                                        -0.0020    0.0000    0.0020    0.0040


WEEKDAY = 2  subtracted from:
WEEKDAY     Lower    Center     Upper  ---+---------+---------+---------+---
3        -0.000285  0.001466  0.003217                (-------*--------)
4        -0.000464  0.001298  0.003061               (-------*--------)
5        -0.001175  0.000590  0.002355           (--------*--------)
                                        ---+---------+---------+---------+---
                                        -0.0020    0.0000    0.0020    0.0040


WEEKDAY = 3  subtracted from:
WEEKDAY     Lower    Center     Upper  ---+---------+---------+---------+---
4        -0.001930 -0.000167  0.001595     (--------*--------)
5        -0.002640 -0.000876  0.000888  (--------*-------)
                                        ---+---------+---------+---------+---
                                        -0.0020    0.0000    0.0020    0.0040

WEEKDAY = 4  subtracted from:
WEEKDAY     Lower    Center     Upper  ---+---------+---------+---------+---
5        -0.002481 -0.000708  0.001064  (-------*--------)
                                        ---+---------+---------+---------+---
                                        -0.0020    0.0000    0.0020    0.0040
```

Bonferroni Simultaneous Tests
Response Variable intradayROR
All Pairwise Comparisons among Levels of WEEKDAY

```
WEEKDAY = 1  subtracted from:
         Difference       SE of             Adjusted
WEEKDAY   of Means    Difference  t-Value    p-Value
2         0.000228     0.000638    0.3579     1.0000
3         0.001694     0.000638    2.6570     0.0793
4         0.001527     0.000641    2.3822     0.1727
5         0.000818     0.000642    1.2751     1.0000


WEEKDAY = 2  subtracted from:
         Difference       SE of             Adjusted
WEEKDAY   of Means    Difference  t-Value    p-Value
3         0.001466     0.000623    2.3513     0.1877
4         0.001298     0.000628    2.0691     0.3862
```

```
5              0.000590     0.000628    0.9390     1.0000

WEEKDAY = 3  subtracted from:
          Difference        SE of                  Adjusted
WEEKDAY   of Means      Difference   t-Value       p-Value
4         -0.000167     0.000627     -0.267        1.000
5         -0.000876     0.000628     -1.395        1.000

WEEKDAY = 4  subtracted from:
          Difference        SE of                  Adjusted
WEEKDAY   of Means      Difference   T-Value       p-Value
5         -0.000708     0.000631     -1.123        1.000
```

Trading strategy based on day-of-the-week and yesterday's intraday ROR. At the closing bell, the investor will obtain the intraday ROR for the day, Q_t. The investor will execute a day trade on the next day if its predicted intraday ROR for tomorrow is large enough, that is, if $\hat{Q}_{t+1} > c$. A decision risk analysis to determine c proceeds as follows, comparing the net expected ROR of trading tomorrow with that of not trading tomorrow. It is assumed that tomorrow's daily interest rate IR_{t+1} is known. The ROR of not trading tomorrow is IR_{t+1}. The predicted net ROR of trading tomorrow is $\hat{Q}_{t+1} - TC$, where TC is the transaction cost, expressed as a rate. The difference is $\hat{Q}_{t+1} - TC - IR_{t+1} = \hat{Q}_{t+1} - c$, where $c = TC + IR_{t+1}$. So a strategy is to trade tomorrow if $\hat{Q}_{t+1} > TC + IR_{t+1}$. As mentioned, here TC is expressed as a rate.

The proceeds are at a rate IR_t on days t when the investor sits on the sidelines and at a rate $Q_t - TC$ on days t when the investor trades. The daily proceeds can be added up on a spreadsheet. Again, better estimates of how such a strategy would perform in practice would be obtained by saving the more recent part of the dataset as a test set.

An alternative is to using a rolling computation, updating the parameter estimates. This could be done each day, one the spreadsheet is set up or program is written. The window might be, for example, 252 days (the typical number of trading days in a year), or perhaps 504 days.

8.11   Simultaneous Equations Models

Above it was seen how relationships of price and quantity to lags of one another lead to simultaneous equations and then to an autoregression for price.

More generally, there can be a system of equations involving a number of variables. There are large-scale econometric models of the U.S. economy, involving a large number of variables.

8.12   Appendix 8A: Growth Rates and Rates of Return

If a principal amount P_0 earns interest for one period at a rate r per period, the principal at the end of the period is $P_1 = P_0(1 + r)$. If this continues for n periods, then $P_n = P_0(1 + r)^n$. If there are different rates $r_1, r_2, \ldots, r_n$ in the different periods, then this is $P_n = P_0(1 + r_1)(1 + r_2) \ldots (1 + r_n)$.

8.12.1   Compound Interest

If there is compounding m times per period, for n periods, with rates $R_1, R_2, \ldots, R_n$, then $P_n = P_0(1 + R_1/m)^m(1 + R_2/m)^m \ldots (1 + r_n/m)^m$. If the rate R is compounded m times during the period, then $P_1 = P_0(1+r/m)^m$.

To convert an annual rate R to a monthly rate, one can approximate as $R/12$. However, a more precise computation of the monthly rate is $(1 + R)^{1/12} - 1$. For example, if $R = .06$ or 6% per year, the monthly rate is approximately $0.06/12 = .005$ or 0.5% per month, but on a compound basis the rate is $1.06^{1/12} - 1 = .0048675$ or 0.45675% per month, a little less than 0.5% per month.

If a rate r is compounded continuously over the period, then

$$P_1 = \lim_{m \to \infty} (1 + r/m)^m = P_0 e^r.$$

If the compounding is continuous, then

$$P_n = P_0 e^{r_1} e^{r_2} \ldots e^{r_n} = P_0 \exp(r_1 + r_2 + \ldots + r_n) = P_0 \exp\left(\sum_{t=1}^{n} r_t\right).$$

8.12.2   Geometric Brownian Motion

If the rates of return $r_1, r_2, \ldots, r_n$ are random variables, then $\{P_n, n = 1, 2, \ldots\}$ is a random process. If in addition the r_t are i.i.d. (independent and identically distributed) according to a Normal distribution, then the process $P_n = P_0 \exp(\sum_{t=1}^{n} r_t)$ is called *geometric (exponential) Brownian motion*, or GBM.

Note that in the GBM, the variables r_t are the continuous RORs, $r_t = \ln P_t - \ln P_{t-1}, t = 1, 2, \ldots, n$, and $\ln P_n = \ln P_0 \exp(\sum_{t=1}^{n} r_t) = \ln P_0 + \ln(\exp(\sum_{t=1}^{n} r_t)) = \ln P_0 + \sum_{t=1}^{n} r_t = \ln P_0 + \sum_{t=1}^{n} (\ln P_t - \ln P_{t-1})$. Taking the term $\ln P_0$ to the left-hand side, we have

$$\ln P_n - \ln P_0 = \sum_{t=1}^{n} (\ln P_t - \ln P_{t-1}) = \sum_{t=1}^{n} r_t.$$

This says that the continuous ROR for the whole period is the sum of the n RORs. For example, weekly continuous ROR is the sum of the daily RORs. This property is not shared by ordinary (discrete) ROR, $R_t = (P_t - P_{t-1})/P_{t-1}$. For example, let daily prices be P_0 for last Friday, P_1 for this Monday, P_2 for this Tuesday, P_3 for this Wednesday, P_4 for this Thursday, and P_5 for this Friday. Then the weekly ROR is $(P_5 - P_0)/P_0$, and in terms of the daily RORs $R_t = (P_t - P_{t-1})/P_{t-1}$, $t = 1, 2, 3, 4, 5$, we have

$$
\begin{aligned}
(P_5 - P_0)/P_0 &= [(P_5 - P_4) + (P_4 - P_3) + (P_3 - P_2) + (P_2 - P1) + (P_1 - P_0)]/P_0 \\
&= [P_4(P_5 - P_4)/P_4 + P_3(P_4 - P_3)/P_3 + P_2(P_3 - P_2)/P_2 \\
&\quad + P_1(P_2 - P1)/P_1 + P_0(P_1 - P_0)/P_0]/P_0 \\
&= [P_4\,R_5 + P_3\,R_4 + P_2\,R_3 + P_1\,R_2 + P_0\,R_1]/P_0 \\
&= (P_4/P_0)R_5 + (P_3/P_0)R_4 + (P_2/P_0)R_3 + (P_1/P_0)R_2 \\
&\quad + (P_0/P_0)P_0 R_1 \\
&\neq R_1 + R_2 + R_3 + R_4 + R_5.
\end{aligned}
$$

Thus, the weekly ROR is not a simple sum but rather a *weighted* sum of the daily RORs in which the weights P_t/P_0 depend on the daily prices P_t.

8.12.3 Average Rates of Return

If annual rates of increase are $r_1, r_2, \ldots, r_n$, then the average rate of increase r is derived from

$$(1+r)^n = (1+r_1) \times (1+r_2) \times \cdots \times (1+r_n).$$

This means that it is given by the formula

$$r = [(1+r_1) \times (1+r_2) \times \cdots \times (1+r_n)]^{1/n} - 1.$$

If instead of the rates of increase the annual sales sales $y_0, y_1, y_2, \ldots, y_n$ are given, then the average annual rate of increase r is defined by

$$y_n = (1+r)^n y_0.$$

This gives $r = (y_n/y_0)^{1/n} - 1$. Suppose sales in eight years a company's capitalization grew from 2 m\$. to 10 M\$. What is the average annual rate of growth, r? The solution is given by $y_n = y_8 = 10\text{M\$} = (1+r)^n y_1 = (1 + r)^8\,(2\text{M\$})$, $(1+r)^8 = 5$, $1+r = 5^{1/8} \approx 1.223$, $r \approx .223$ or 22.3% per year.

8.12.4 Section Exercises: Exponential and Log Functions

8.1 $\ln(e^x) = x$. Why?
Solution: The two functions are inverses of one another, so the answer is x.

8.2 For $x > 0$, $e^{\ln x} = x$. Why?
Solution: The two functions are inverses of one another, so the answer is x.

8.3 $\ln(2e) = 1 + \ln(2)$. Why? *Solution:* $\ln(2e) = \ln(2) + \ln(e) = \ln(2) + 1.$,

8.4 Which of the following is closest to the value of the number $1/e$: $1/3$, 0.3679, $1/2$, 3, or 3.14159? *Solution:* The number e is a little less than 3, so $1/e$ is approximately $1/3$. More precisely, $1/e = e^{-1} \approx 1/2.71818 \approx 0.3679$.

8.13   Appendix 8B: Prediction after Data Transformation

8.13.1   Prediction

Because the conditional expectation minimizes the mean squared error of prediction, the predicted value of Y_{t+1} is usually taken as the estimate of the conditional expectation of Y_{t+1}, given the past data, that is, the past history of the time series, up to and including time t, denoted by $\mathcal{H}_t$. (The past history $\mathcal{H}_t$ includes $y_t, a_t, y_{t-1}, a_{t-1}, \dots$.)

The prediction would be $\hat{Y}_{t+1} = \mathcal{E}[Y_{t+1} \mid \mathcal{H}_t]$. Now, this will involve parameters, which must be estimated from the data. For example, in the case of AR(1), $\mathcal{E}[Y_{t+1} \mid \mathcal{H}_t] = \phi_0 + \phi_1 y_t$, and ϕ_0 and ϕ_1 must be estimated. So the prediction $\hat{Y}_{t+1}$ will be $\hat{\phi}_0 + \hat{\phi}_1 y_t$.

8.13.2   Prediction after Transformation

But suppose that the data have been transformed, according to $Z = h(Y)$. Then the model is built in terms of Z. Then how should Y be predicted? The inverse transform is $Y = h^{-1}(Z) = g(Z)$, where we let $g(\cdot) = h^{-1}(\cdot)$, for short. One way to proceed would be simply to plug in $\hat{Z}$, taking the prediction of Y to be $\hat{Y} = g(\hat{Z})$.

For example, suppose $Z = \ln Y$ and the model is AR(1). Then $\hat{Z}_{t+1} = \hat{\phi}_0 + \hat{\phi}_1 z_t$. The inverse transform is $Y = h^{-1}(Z) = g(Z) = e^Z$ and a prediction of Y_{t+1} is $\exp(\hat{Z}_{t+1}) = \exp(\hat{\phi}_0 + \hat{\phi}_1 z_t) = \exp(\hat{\phi}_0) \exp(\hat{\phi}_1 z_t) = \exp(\hat{\phi}_0) \exp(z_t)^{\hat{\phi}_1} = \exp(\hat{\phi}_0) y_t^{\hat{\phi}_1}$.

8.13.3   Unbiasing

This is not an unreasonable way to proceed. But the predicted value of Y_{t+1} is supposed to be its conditional expectation given the data, and $\mathcal{E}[Y] = \mathcal{E}[h^{-1}(Z)] \neq h^{-1}(\mathcal{E}[Z])$, unless $h(\cdot)$ is linear.

An approach to finding an unbiasing term proceeds as follows. Let $h^{-1}(z) = g((z)$ for short. The first two terms of a Taylor series approximation, expanding $g(z)$ about the point z_0, are

$$y = g(z) \approx g(z_0 + (z - z_0)g'(z_0) + (1/2)(z - z_0)^2 g''(z_0).$$

Taking z_0 to be μ_z, here representing $\mathcal{E}[Z \mid \mathcal{H}_t]$, gives

$$y = g(z) \approx g(\mu_z) + (z - \mu_z)g'(\mu_z) + (1/2)(z - \mu_z)^2 g''(\mu_z).$$

This gives

$$
\begin{aligned}
\mathcal{E}[Y] &= \mathcal{E}[g(Z)] \\
&\approx g(\mu_z) + \mathcal{E}[(Z - \mu_z)\,g'(\mu_z)] + (1/2)\mathcal{E}(Z - \mu_z)^2 g''(\mu_z) \\
&= g(\mu_z) + 0 + (1/2)\sigma_{z_x}^2 \, g''(\mu_z) \\
&= g(\mu_z) + (1/2)\sigma_z^2 \, g''(\mu_z).
\end{aligned}
$$

So the bias correction term added to $g(\mu_z)$ is $(1/2)\sigma_z^2 \, g''(\mu_z)$.

8.13.4   Application to the Log Transform

The log transform is $Z = h(Y) = \ln Y$. The inverse is $Y = g(Z) = \exp(Z)$. The derivatives involved are $g'(z) = \exp(z)$ and $g''(z) = \exp(z)$. The unbiasing term is $(1/2)\,\sigma_z^2\, g''(\mu_z) = (1/2)\,\sigma_z^2 \exp(\mu_z)$. Then μ_y is approximated by $g(\mu_z) + (1/2)\sigma_z^2 \, g''(\mu_z) = \exp(\mu_z) + (1/2)\,\sigma_z^2 \exp(\mu_z) = \exp(\mu_z)\,[1 + (1/2)\,\sigma_z^2]$. That means the bias-corrected prediction is taken as $\hat{Y}_{t+1} = \exp(\hat{Z}_{t+1})\,[1 + (1/2)\,s_z^2]$, where s_z^2 is the mean squared error obtained in fitting the model to $\{Z_t\}$. Note that the bias-corrected prediction is the plug-in prediction $\exp(\hat{Z}_{t+1})$ times a bias-correction factor $[1 + (1/2)\,s_z^2]$.

8.13.5   Generalized Linear Models

Generalized linear models generalize the multiple linear regression model given by the regression function $\mathcal{E}[Y \mid x_1, x_2, \ldots, x_p] = \beta_0 + \beta_1 x_1 + \beta_2 x_2 + \cdots + \beta_p x_p$. A transform $h(\mathcal{E}[Y \mid x_1, x_2, \ldots, x_p])$ is taken as obeying a multiple linear regression model: $h(\mathcal{E}[Y \mid x_1, x_2, \ldots, x_p]) = \beta_0 + \beta_1 x_1 + \beta_2 x_2 + \cdots + \beta_p x_p$. Thus $\mathcal{E}[Y \mid x_1, x_2, \ldots, x_p] = h^{-1}(\beta_0 + \beta_1 x_1 + \beta_2 x_2 + \cdots + \beta_p x_p) = g(\beta_0 + \beta_1 x_1 + \beta_2 x_2 + \cdots + \beta_p x_p)$, and the preceding analysis can be applied to improve upon the plug-in prediction $g(b_0 + b_1 x_1 + b_2 x_2 + \cdots + b_p x_p)$.

   A word on terminology. Do not confuse the terms *Generalized* Linear Model and *General* Linear Model. *Generalized Linear Model* refers to a model where the transform of the mean of Y obeys a multiple linear regression model. The term *General Linear Model* refers to a model where Y itself obeys a multiple linear regression model but allows that model to be other than a standard simple linear regression or balanced ANOVA.

8.14 Appendix 8C: Representation of Time Series

Representation of time series includes expressing Y_t in terms of past values $y_{t-1}, y_{t-2}, \ldots$ and ε_t or in terms of the error sequence $\{\varepsilon_t\}$. Such representations can be used to obtain means, variances, and covariances, and to obtain predicting formulas. Some simple results are stated and derived. For more advanced results and details, we defer to books on time series analysis *per se*.

8.14.1 Operators

Throughout, the operators $\mathcal{E}$ for mathematical expectation, $\mathcal{V}$ for variance, and $\mathcal{C}$ for covariance are used.

In this section, some additional operators are used. The *backshift operator* $\mathcal{B}$ operates as $\mathcal{B}[Y_t] = Y_{t-1}$. The *identity operator* $\mathcal{I}$ is $\mathcal{I}[Y_t] = Y_t$. Note that the difference operator $\mathcal{D}$ is $\mathcal{I} - \mathcal{B}$, that is,

$$\mathcal{D}[Y_t] = (\mathcal{I} - \mathcal{B})[Y_t] = \mathcal{I}[Y_t] - \mathcal{B}[Y_t] = Y_t - Y_{t-1}.$$

The backshift operator is also called the *lag operator* and written $\mathcal{L}[Y_t] = Y_{t-1}$.

We denote operators by script letters so that they are not confused notationally with scalars, vectors, or matrices, and so that it is clear that operators are different than ordinary functions.

8.14.2 White Noise

The error sequence $\{\varepsilon_t\}$ is *white noise*. That means $\mathcal{E}[\varepsilon_t] = 0$, $\mathcal{V}[\epsilon_t] = \sigma_\varepsilon^2$, a constant not depending on t, and $\mathcal{C}[\varepsilon_t, \varepsilon_u] = 0$, for $t \neq u$. A short way of writing this is $\mathcal{C}[\varepsilon_t, \varepsilon_u] = \sigma_\varepsilon^2 \delta_{tu}$, where δ_{tu} is the *Kronecker delta*: $\delta_{tu} = 1$ if $t = u$ and $\delta_{tu} = 0$ if $t \neq u$.

8.14.3 Stationarity

Strong stationarity means that all the finite-dimensional distributions are time-invariant; that is, for any p and any time points $(t_1, t_2, \ldots, t_p)$, the joint distribution of $Y_{t_1+h}, Y_{t_2+h}, \ldots, Y_{t_p+h}$ does not depend upon h. For time series analysis, *weak stationarity (covariance stationarity)* suffices. This means that $\mathcal{E}[Y_t]$ does not depend upon t, and for every $k = 0, 1, 2, \ldots$, $\mathcal{C}[Y_t, Y_{t-k}]$ depends only upon k and not t. In what follows it is assumed that the series $\{Y_t\}$ is covariance stationary. In particular, its mean, variance, and covariances will not depend upon t.

8.14.4   AR

The AR(1) model is $Y_t = \phi_0 + \phi_1 y_{t-1} + \varepsilon_t$. The mean can be found as follows: $\mu = \mathcal{E}[Y_t] = \mathcal{E}[\phi_0 + \phi_1 Y_{t-1}] = \phi_0 + \phi_1 \mu$; this gives

$$\mu - \phi_1 \mu = (1 - \phi_1)\mu = \phi_0,$$

so $\mu = \phi_0/(1 - \phi_1)$. The deviation from the mean is $\tilde{Y}_t = Y_t - \mu$. Its conditional expectation is $\phi \tilde{y}_{t-1}$, because

$$
\begin{aligned}
\mathcal{E}[\tilde{Y}_t \mid y_{t-1}] &= \phi_0 - \mu + \phi_1 y_{t-1} \\
&= \phi_0 - \phi_0/(1 - \phi_1) + \phi_1 y_{t-1} = \phi_0 \left[1 - 1/(1 - \phi_1)\right] + \phi_1 y_{t-1} \\
&= \phi_0[-\phi_1/(1 - \phi_1)] + \phi_1 y_{t-1} = [\phi_0/(1 - \phi_1)](-\phi_1) + \phi_1 y_{t-1} \\
&= -\phi_1 \mu + \phi_1 y_{t-1} = \phi_1 (y_{t-1} - \mu) = \phi_1 \tilde{y}_{t-1}.
\end{aligned}
$$

The model can be rewritten as

$$\tilde{Y}_t - \phi_1 \tilde{y}_{t-1} = \varepsilon_t, \text{ or } (\mathcal{I} - \phi_1 \mathcal{B})[\tilde{Y}_t] = \varepsilon_t.$$

Proceeding formally to write the inverse operator gives

$$\tilde{Y}_t = (\mathcal{I} - \phi_1 \mathcal{B})^{-1}[\varepsilon_t],$$

where, given a sequence $x_1, x_2, \ldots, x_n, \ldots$, the *back-shift operator* $\mathcal{B}$ is $\mathcal{B}[x_t] = x_{t-1}$. The expression involving the inverse can be expanded using the formula for the sum of a geometric series with common ratio r:

$$1 + r + r^2 + \cdots = \frac{1}{1 - r}, \text{ for } |r| < 1.$$

Take $r = \phi_1 \mathcal{B}$ to obtain a representation of $\tilde{Y}_t$ in terms of $\{\varepsilon_t\}$. Stationarity implies $|\phi_1| < 1$. The representation is

$$\tilde{Y}_t = \sum_{k=0}^{\infty} \phi_1^k \varepsilon_{t-k}.$$

8.14.4.1   Variance

The variance is

$$\mathcal{V}[Y_t] = \mathcal{V}[\tilde{Y}_t] = \mathcal{V}[\sum_{k=0}^{\infty} \phi_1^k \varepsilon_{t-k}] = \sigma_\varepsilon^2 \sum_{k=0}^{\infty} \phi_1^{2k} = \frac{\sigma_\varepsilon^2}{1 - \phi_1^2},$$

where the fact that the covariances $\mathcal{C}[\varepsilon_t, \varepsilon_u]$ for $t \neq u$ are zero has been used.

8.14.4.2 Covariances and Correlations

The notation Y_t denotes the value of Y at time t. For $k = 1, 2, \ldots,$ the notation Y_{t-k} denotes the value of Y at time $t - k$, that is, k time units earlier. The integer k is called the *lag*. The lag-k covariance is

$$
\begin{aligned}
\mathcal{C}[\, Y_t, Y_{t-k}\,] &= \mathcal{C}[\, \tilde{Y}_t, \tilde{Y}_{t-k}\,] \\
&= \mathcal{C}[\, \sum_{i=0}^{\infty} \phi_1^i\, \varepsilon_{t-i}, \ \sum_{j=0}^{\infty} \phi_1^j\, \varepsilon_{t-k-j}\,] \\
&= \mathcal{C}[\, \sum_{i=0}^{k-1} \phi_1^i\, \varepsilon_{t-i} + \sum_{i=k}^{\infty} \phi_1^i\, \varepsilon_{t-i}, \ \sum_{j=0}^{\infty} \phi_1^j\, \varepsilon_{t-k-j}\,] \\
&= \mathcal{C}[\, \sum_{i=0}^{k-1} \phi_1^i\, \varepsilon_{t-i}, \sum_{j=0}^{\infty} \phi_1^j\, \varepsilon_{t-k-j}\,] \\
&\quad + \mathcal{C}[\, \sum_{i=k}^{\infty} \phi_1^i\, \varepsilon_{t-i}, \ \sum_{j=0}^{\infty} \phi_1^j\, \varepsilon_{t-k-j}\,] \\
&= 0 + \mathcal{C}[\, \sum_{i=k}^{\infty} \phi_1^i\, \varepsilon_{t-i}, \ \sum_{j=0}^{\infty} \phi_1^j\, \varepsilon_{t-k-j}\,] \\
&= \mathcal{C}[\, \sum_{i=k}^{\infty} \phi_1^i\, \varepsilon_{t-i}, \ \sum_{j=0}^{\infty} \phi_1^j\, \varepsilon_{t-k-j}\,] \\
&= \mathcal{C}[\, \sum_{m=0}^{\infty} \phi_1^{m+k}\, \varepsilon_{t-k-m}, \ \sum_{j=0}^{\infty} \phi_1^{m+k}\, \varepsilon_{t-k-j}\,] \\
&= \mathcal{C}[\, \phi_1^k \sum_{m=0}^{\infty} \phi_1^m\, \varepsilon_{t-k-m}, \ \sum_{j=0}^{\infty} \phi_1^j\, \varepsilon_{t-k-j}\,] \\
&= \phi_1^k\, \mathcal{C}[\, \sum_{m=0}^{\infty} \phi_1^m\, \varepsilon_{t-k-m}, \ \sum_{j=0}^{\infty} \phi_1^j\, \varepsilon_{t-k-j}\,] \\
&= \phi_1^k\, \mathcal{C}[\, \tilde{Y}_{t-k}, \tilde{Y}_{t-k}] = \phi_1^k\, \mathcal{V}[\, \tilde{Y}_{t-k}\,] = \phi_1^k\, \mathcal{V}[\, Y_{t-k}\,] \\
&= \phi_1^k\, \sigma_\varepsilon^2 / (1 - \phi_1^2).
\end{aligned}
$$

The lag-k autocorrelation is $\phi^k / (1 - \phi_1^2)$.

8.14.4.3 Higher-Order AR

Similar calculations yield results for AR(p). Details are omitted here; see one or another of the books on time series in the Bibliography.

8.14.5 MA

The MA(q) model is $Y_t = \mu + \varepsilon_t - \theta_1\varepsilon_{t-1} - \cdots - \theta_q\varepsilon_{t-q}$, where the error sequence $\{\varepsilon_t\}$ is white noise. The mean is $\mathcal{E}[Y_t] = \mu$.

8.14.5.1 Variance

The model $Y_t = \mu + \varepsilon_t - \theta\varepsilon_{t-1}$, where $\{\varepsilon_t\}$ is white noise, is an MA(1) model. The variance $\sigma_y^2 = \mathcal{V}[Y_t]$ is

$$
\begin{aligned}
\mathcal{V}[Y_t] &= \mathcal{V}[\varepsilon_t - \theta\varepsilon_{t-1}] \\
&= \mathcal{V}[\varepsilon_t] + 2\mathcal{C}[\varepsilon_t, -\theta\varepsilon_{t-1}] + \mathcal{V}[-\theta\varepsilon_{t-1}] \\
&= \mathcal{V}[\varepsilon_t] - 2\theta\mathcal{C}[\varepsilon_t, \varepsilon_{t-1}] + (-\theta)^2\mathcal{V}[\varepsilon_{t-1}] \\
&= \sigma_\varepsilon^2 + 0 + \theta^2\sigma_\varepsilon^2 \\
&= (1+\theta^2)\sigma_\varepsilon^2.
\end{aligned}
$$

The variance in an MA(q) model can be obtained as $\sigma_\varepsilon^2(1+\theta_1^2+\theta_2^2+\cdots+\theta_q^2)$ by similar operations.

8.14.5.2 Correlation

In an MA(1), Y_t and Y_{t-1} are correlated but Y_t and Y_{t-k} for $k = 2, 3, \ldots$ are uncorrelated. In an MA(q), observations within q time periods of one another are correlated; observations farther apart are not.

8.14.5.3 Representing the Error Variables in Terms of the Observations

As mentioned in the text, in developing forecasts for models with MA parts, it is necessary to represent the realized errors e_t in terms of the values of past and current observations $y_t, y_{t-1}, \ldots$. In terms of operators $\mathcal{B}[Y_t]$, $=$ Y_{t-1} and $\mathcal{I}$, the identity operator $\mathcal{I}[Y_t] = Y_t$, write $Y_t = \mathcal{I}\varepsilon_t - \theta\mathcal{B}\varepsilon_t = [\mathcal{I} - \theta\mathcal{B}]\varepsilon_t$. Proceeding formally, write $[\mathcal{I} - \theta\mathcal{B}]^{-1}Y_t = \varepsilon_t$, $\varepsilon_t = [\mathcal{I} + \theta\mathcal{B} + \theta^2\mathcal{B}^2 + \cdots]Y_t = Y_t + \theta Y_{t-1} + \theta^2 Y_{t-2} + \cdots$. The realized value is $e_t = y_t + \theta y_{t-1} + \theta^2 y_{t-2} + \cdots$. Now, the data series goes back only to y_1. Note, however, that $e_t \approx y_t - \theta y_{t-1} + \theta^2 Y_{t-2} + \cdots + \theta^{t-1}y_1$. The approximation is good because higher powers of θ are very small.

As an example, consider predicting the next observation in an MA(1). The model is $Y_{t+1} = \varepsilon_{t+1} - \theta\varepsilon_t$. The conditional expectation is

$$
\begin{aligned}
\mathcal{E}[Y_{t+1}\,|\,\mathcal{H}_t] &= \mathcal{E}[\varepsilon_{t+1} - \theta\varepsilon_t\,|\,\mathcal{H}_t] \\
&= \mathcal{E}[\varepsilon_{t+1}\,|\,\mathcal{H}_t] - \theta\mathcal{E}[\varepsilon_t|\mathcal{H}_t] \\
&= 0 - \theta e_t = -\theta(y_t + \theta y_{t-1} + \theta^2 y_{t-2} + \cdots) \\
&\approx -\theta(y_t + \theta y_{t-1} + \theta^2 y_{t-2} + \cdots + \theta^{t-1}y_1) \\
&= -\theta y_t - \theta^2 y_{t-1} - \cdots - \theta^t y_1.
\end{aligned}
$$

Note, then, that for MA(1), $\hat{Y}_{t+1} = -\theta(y_t - \theta y_{t-1} - \theta y_{t-2} - \cdots = -\theta(y_t -$

$\hat{Y}_t$). That is, the predicted value is $-\theta$ times the preceding prediction error $y_t - \hat{Y}_t$. Of course, in these formulas, an estimate $\hat{\theta}$ would be plugged in for θ.

The method of representation of Y_t in terms of $\{\varepsilon_t\}$ can be extended to MA(q) by inverting the operator $\mathcal{I} - \theta_1 \mathcal{B} - \theta_2 \mathcal{B}^2 + \cdots + \theta_q \mathcal{B}^q$. This involves finding roots of the corresponding auxiliary polynomial equation.

8.14.6   ARMA

ARMA models

$$Y_t = \theta_0 + \phi_1 y_{t-1} + \cdots + \phi_p y_{t-1} + \varepsilon_t - \theta_1 \varepsilon_{t-1} - \cdots - \theta_q \varepsilon_{t-q}$$

can be similarly represented. We defer to advanced books on time series for details.

8.15   Summary

Temporal data are data observed in time.   *Time series* are temporal data observed at regular time intervals, for example, daily, weekly, monthly, quarterly, or annually.

A *trend* is a general up movement or down movement.

Moving averages smooth the data and facilitate the detection of trends in the data. The average used may be the mean or the median. The window *width* is the number of observations averaged. A moving average may be *centered,* or may work back in time.

The smoothed value S_t may be taken as a prediction $\hat{Y}_{t+1}$ of the next value Y_{t+1}.

Often, the changes, or *differences*, rather than the original values are analyzed. That is, a series may be pre-processed by differencing to yield a level series for analysis. Sometimes a second difference may be used. The first difference corresponds to velocity; the second, to acceleration.

Exponentially weighted moving averages weight more recent observations more heavily. The *smoothing constant* is the weight given to the most recent observation.

Autoregressive models seek to explain and predict a response variable in terms of its past values.

ARIMA models may include differencing, autoregressive terms and moving average terms. The moving average part of the model is autoregression in the errors.

Dynamic regression models (*transfer function models*) try to explain and predict a response variable in terms of past values of other variables.

Monthly and quarterly data are examples of *seasonal data.*

If in a spreadsheet the rows correspond to years and the columns to quarters, then the row percents give the *seasonal pattern.* If the seasonal pattern is stable, it can be used for purposes of forecasting.

8.16 Chapter Exercises

8.16.1 Applied Exercises

8.5 Given the time series $y_1 = 218, y_2 = 217, y_3 = 214, y_4 = 216, y_5 = 211, y_6 = 214, y_7 = 212$, compute the running median of three.
Solution: Let rMdn denote the running median. The computation proceeds as $\text{rMdn}_2 = \text{median}\{218, 217, 214\} = 217$, $\text{rMdn}_3 = \text{median}\{217, 214, 216\} = 216$. Continue the computation, obtaining smoothed values 214, 214, and 212.

8.6 (continuation) Plot the original and smoothed values on the same graph. Note that the plot of the rMdn sequence will be smoother than that of the original sequence. Note the consistent downtrend of the smoothed series, despite the slight ups and downs of the original series.

8.7 (continuation) Compute the predictor-corrector formula for a six-point sum-of-digits WMA.

8.8 Show that if $\alpha = 2/(n+1)$, then $n = (2 - \alpha)/\alpha = 2/\alpha - 1$.

8.9 Weights for an EMA. Compute the EMA weights for $\alpha = .1$ and compare them with the weights $1/n$ for a moving average, where $n = (2 - .1)/.1 = 19$. *Remark.* The weights for the MA are then $1/19$, or just above .05.

8.10 Weights for an EMA. Compute the EMA weights for $\alpha = .2$ and compare them with the weights $1/n$ for a moving average, where $n = (2 - .2)/.2 = 9$. *Remark.* The weights for the MA are then $1/9 \approx .111$, or just above .10.

8.11 Stable seasonal pattern. Suppose sales in 2009 were 5.0 M\$, and there is a stable seasonal pattern across quarters, namely 10%, 20%, 30%, and 40% for the first, second, third, and fourth quarters, respectively. Using the predictor that total sales for 2010 will be 10% higher than sales were for 2009, what is the forecast of total sales for 2010? *Solution:* 5.0 M\$ $\times$ 1.10 = 5.5 M\$.

8.12 (continuation) Using this forecast of annual sales, predict sales for the second quarter of 2010 using the stable seasonal pattern. *Solution:* The forecast of second quarter sales is 20% of 5.5 M\$ = 1.1 M\$.

8.13 Update the Best Buy quarterly sales data in the text (Table 8.4. Update the analyses using the augmented dataset.

8.14 Suppose $Y_t = \phi_0 + \phi_1 y_{t-1} + \varepsilon_t$, $t = 1, 2, \ldots, n$. If the one-step-ahead predictor is $\hat{Y}_{n+1} = \phi_0 + \phi_1 y_n$, and if the two-step-ahead predictor $\hat{Y}_{n+2}$ is taken to be $\phi_0 + \phi_1 \hat{Y}_{n+1}$, what is $\hat{Y}_{n+2}$ in terms of y_n?

8.15 (continuation) Given observations $y_1, y_2, \ldots, y_n$ of a time series, denote the prediction of Y_{t+h} at time n that is, the h-step ahead prediction at time n, by $\hat{Y}_n(h)$, $h = 1, 2, \ldots$. Use the method of the preceding problem to compute this for AR(1).

8.16.2   Mathematical Exercises

8.16 Add up $1 + 2 + 3 + 4 + 5 + 6$ as $(1 + 6) + (2 + 5) + (3 + 4)$.

8.17 Add up $1 + 2 + 3 + 4 + 5$ as $(1+5) + (2+4) + 3$.

8.18 (continuation) Compute the predictor-corrector formula for a five-point sum-of-digits WMA.

8.19 Show that $1 + 2 + 3 + \cdots + n = n(n+1)/2$.

8.20 Compute the correlation between $(Y_1+Y_2+Y_3)/3$ and $(Y_4+Y_5+Y_6)/3$ when the Y_t are uncorrelated with common variance σ^2.

8.21 Compute the correlation between $(Y_1+Y_2+Y_3)/3$ and $(Y_4+Y_5+Y_6)/3$ when $Y_t = \varepsilon_t - \theta\,\varepsilon_{t-1}$ and $\{\varepsilon_t\}$ is white noise (uncorrelated with common variance σ^2).

8.22 Compute the correlation between $(Y_1+Y_2+Y_3)/3$ and $(Y_4+Y_5+Y_6)/3$ when $Y_t = \phi\,Y_{t-1} + \varepsilon_t$, where $\{\varepsilon_t\}$ is white noise and Y_0 is uncorrelated with $\{\varepsilon_t, Y_t, t = 1, 2, \ldots\}$ and has variance σ^2.

8.23 For the preliminary estimate of θ in an MA(1) model, set up the indicated quadratic equation in a form so that the constant is 1.

8.24 (continuation) Show that this implies that the product of the roots is 1, that is, that the roots are reciprocal.

8.25 (continuation) Solve the equation.

8.26 Quarterly effects.   Write a model for quarterly effects with monthly data. Show how it is a restriction of the model for monthly effects. (There will be four quarterly parameters rather than twelve monthly parameters but each is repeated three times, because the data are monthly. Also, the number of dummy variables required in either case is one less than the number of categories.)

8.27 (continuation)   Show how to test the adequacy of quarterly effects compared to monthly effects.

8.28 Estimation.   Show that the mean squared error $\int (y - a)^2 f_Y(y)\,dy$ is minimized by $a = \mu_y$.

8.29 Prediction.   Show that the mean squared error

$$\int [y - T(x)]^2 f_{Y|X}(y|x)\,dy$$

is minimized by $T(x) = \mathcal{E}[Y \mid x]$.

8.30 Prediction.   Show that the mean squared error

$$\int [y_{t+1} - T(y_t)]^2 f_{Y_{t+1}|Y_t}(y_{t+1}|y_t)\,dy_t$$

is minimized over predictors $T(y_t)$ by $T(y_t) = \mathcal{E}[Y_{t+1} \mid y_t]$.

8.17   Bibliography

Anderson, T. W. (1994).   *The Statistical Analysis of Time Series.* Wiley, New York. Wiley Classics Library Edition published 1994. (A time-series text—more advanced.)

Annenberg Foundation (1989). *Against All Odds: Inside Statistics.* David Moore, Content Developer. Annenberg/Corporation for Public Broadcasting (CPB) Project. Consortium for Mathematics and Its Applications (COMAP) and Chedd-Angier. www.learner.org/resources/series65.html. (Videos)

Box, George E. P., Jenkins, Gwilym, and Reinsel, Greg (1994).   *Time Series Analysis and Forecasting. 2nd ed.* Prentice Hall, Upper Saddle River, NJ. (A time-series text—more advanced.)

Box, George E. P., and Jenkins, Gwilym (1970).   *Time Series Analysis and Forecasting.* Holden-Day, San Francisco, CA. rev. ed. 1976. (A time-series text—more advanced.)

Chen, Rong, and Fomby, Thomas (1999).   Forecasting with stable seasonal pattern models with an application of Hawaiian tourist data. *Journal of*

Business and Economic Statistics, **17,** 497–504. (Article)

Cryer, Jonathan (1986). *Time Series Analysis.* Duxbury Press, Boston, MA. (A time-series text—intermediate.)

Enders, Walter (2004). *Applied Econometric Time Series Analysis. 2nd ed.* Wiley, New York. (A time-series text—more advanced.)

Liu, Lon-Mu (2008). *Time Series Analysis and Forecasting. 2nd ed.* SCA (Scientific Computing Associates), Villa Park, IL. (A time-series text—more advanced.)

McCleary, R., and Hay, R. A., Jr. (1980). *Applied Time Series Analysis for the Social Sciences.* Sage Publications, Beverly Hills, CA. (A time-series text–introductory.)

Nelson, C. R., (1973). *Applied Time Series Analysis for Managerial Forecasting.* Holden-Day, San Francisco, CA. (A time-series text—introductory.)

Ortiz Rodriguez, Rosario de los Angeles (2008). A Statistical Look at Day Trading. PhD dissertation, Dept. of Mathematics, Statistics, and Computer Science, University of Illinois at Chicago. (Dissertation)

Parzen, Emanuel (1999). *Stochastic Processes.* Classics in Applied Mathematics, **24.** SIAM: Society for Industrial and Applied Mathematics. Philadelphia, PA. (First published 1962, Holden-Day, San Francisco, CA.) (On stochastic processes in general.)

Ross, Sheldon M. (1996). *Stochastic Processes. 2nd ed.* John Wiley & Sons, New York. (On stochastic processes in general.)

St. John, David (2010). Technical Analysis Based on Moving Average Convergence and Divergence. PhD dissertation, Dept. of Mathematics, Statistics, and Computer Science, University of Illinois at Chicago. (Dissertation)

Stokes, Houston H. (1997). *Specifying and Diagnostically Testing Econometric Models. 2nd ed.* Quorum Books / Greenwood Press, Westport CT. (A time-series text—more advanced.)

Stokes, Houston H., and Neuburger, Hugh H. (1998). *New Methods in Financial Modeling: Explorations and Applications.* Quorum Books / Greenwood Press, Westport, CT. (A time-series text—more advanced.)

Tsay, Ruey S. (2005). *Analysis of Financial time Series. 2nd ed.* John Wiley & Sons, Inc., Hoboken, HJ. (1st ed., 2001.) (A time-series text—more advanced.)

Vandaele, Walter (1983). *Applied Time Series and Box–Jenkins Models.* Academic Press, New York. (A time-series text—intermediate.)

Wei, William W. S. (2004). *Time Series Analysis: Univariate and Multivariate Methods.* Addison Wesley (Pearson), San Francisco, CA. (2nd ed., 2006.) (A time-series text—more advanced.)

8.18 Further Reading

Those interested in further reading on time series analysis might begin with the book by McCleary and Hay (1980) or the book by Nelson (1973). The books on time series in the Bibliography are listed by level of difficulty. (Difficulty is not necessarily a negative attribute, however!) It is good also to put time series in the broader context of stochastic processes. Helpful texts on this subject include Parzen (1962, 1999) and Ross (1996). The Bibliography section above lists books and then other materials (video, articles, and dissertations).

9

Regime Switching Models

CONTENTS

9.1   Introduction

(*Parts of this chapter are more advanced than most of the rest of the book.*)
This chapter discusses *states* or *phases* of the market and the economy.
The states of the market are conventionally called Bull and Bear. The states
of the economy are referred to as *recession, recovery, expansion, contraction.*
Questions arise as to how many such states to use, and whether the nature of
the states detected by statistical methods corresponds to conventional notions
about the phases of the market or the economy.

     The models considered involve a series $\{Y_t, S_t\}$, where Y is observed and
S is the *state*, or *label*. For example, Y_t may be the rate of return of the
S&P500 index in month t, and S_t may be 1 or 0, for Bull or Bear for that
month. One definition of Bull and Bear is that the state variable $S_t = 1$ if

the ROR $Y_t > 0$ and $S_t = 0$ if $Y_t \le 0$. In that case, S_t is directly derivable from the data. In other cases, the labels $\{S_t\}$ are unobserved and then are estimated from the observed process $\{Y_t\}$ by more complicated means. The act of estimating S_t for each t, that is, labeling each time point of a time series with the name of its estimated state, is called *time series segmentation*. A *segment* (*regime, epoch*) is a sequence of time points with the same label.

Some definitions of phases are in terms of patterns; others are obtained in terms of fitting models in which the states are explicit elements of the model, albeit unobservable. Patterns will be considered first, and then hidden state models.

9.2   Bull and Bear Markets

9.2.1   Definitions of Bull and Bear Markets

The popular financial press repeats the conventional wisdom relating to the definition of Bull and Bear markets. One common definition (Vanguard Group; Investopedia) is that a Bear market is indicated by a price decline of 20% or more in a key stock market index from a recent peak over at least a two-month period. Others consider a twelve-month period. With this definition, on the average, the market is in a Bear state every four or five years. During an average Bear market, the S&P loses about 25% of its value. It usually takes eleven to eighteen months for the market to hit bottom. That would be about two or three to four or five quarters, or one to one and a half years.

The simplest definition of Bull and Bear states would be to label a month as a Bear month if the market index went down in that month and as a Bull month if the market index went up in that month. This is the Up-Down (U/D) method considered in Chapter 5. The U/D method definition of a Bull market indicator variable is

$$\text{Bull}_t = [\text{sgm}(x_t) + 1]/2,$$

where x_t is the market ROR in month t and the *sign function* is

$$\text{sgm}(x) = -1 \text{ if } x \le 0 \text{ and } = +1 \text{ if } x > 0.$$

If $\text{sgm}(x_t) = -1$, then $\text{Bull}_t = 0$; and if $\text{sgm}(x_t) = +1$, then $\text{Bull}_t = 1$. Other definitions of Bull/Bear states can involve a moving window of, say, three months of market RORs because some analysts say that three consecutive down months of a market index constitute a Bear market. This notion can be applied to the data considered in Chapter 5. The idea is that the state should remain Bull until there are three consecutive months with negative market ROR and should remain Bear until there are three consecutive months with positive market ROR. Accordingly, define Bull3, a Bull/Bear indicator built on this idea of agreement of the most recent three months. That

TABLE 9.1

Monthly Excess RORs (%) of Market and Fund, with Bull/Bear State Defined as Bull3. (S3 is the moving sum of 3).

	Mo.	Market	Fund	Bull	S3	Bull3
5	J	2.16%	2.45%	1	*	1
yrs	F	−0.32%	−0.50%	1	*	1
ago	M	0.73%	0.75%	1	3	1
	A	0.82%	1.35%	1	3	1
	M	−3.53%	−3.75%	0	2	1
	J	−0.39%	−0.78%	1	2	1
	J	0.09%	0.22%	1	2	1
	A	1.69%	1.96%	1	3	1
	S	2.03%	2.16%	1	3	1
	O	2.69%	3.13%	1	3	1
	N	1.22%	1.31%	1	3	1
	D	0.85%	1.40%	1	3	1
4	J	0.98%	1.14%	1	3	1
yrs	F	−2.63%	−2.99%	0	2	1
ago	M	0.58%	0.71%	1	2	1
	A	3.83%	4.51%	1	2	1
	M	2.81%	3.27%	1	3	1
	J	−2.18%	−1.73%	0	2	1
	J	−3.65%	−4.22%	0	1	1
	A	0.93%	0.94%	1	1	1
	S	3.19%	4.16%	1	2	1
	O	1.15%	1.29%	1	3	1
	N	−4.78%	−3.23%	0	2	1
	D	−1.12%	−0.56%	0	1	1
3	J	−6.54%	−7.32%	0	0	0
yrs	F	−3.71%	−1.97%	0	0	0
ago	M	−0.70%	−1.36%	0	0	0
	A	4.54%	4.38%	1	1	0
	M	0.92%	1.92%	1	2	0
	J	−9.14%	−9.15%	0	2	0

is, take Bull3 to be 0 (Bear) if the past three months had negative market ROR, equal to 1 (Bull) if the past three months had positive market ROR, and equals Bull3_{t-1} otherwise. That is, the Bull/Bear indicator Bull3_t is defined in terms of Bull_t as follows.

Initialization:

$$\text{for } t = 1, 2, 3, \text{Bull3}_t = \text{Bull}_t.$$

Computation:

$$\text{for } t = 4, 5, \ldots, n,$$
$$\text{Bull3}_t = 0 \text{ if } \text{Bull}_t = \text{Bull}_{t-1} = \text{Bull}_{t-2} = 0$$
$$\text{Bull3}_t = 1 \text{ if } \text{Bull}_t = \text{Bull}_{t-1} = \text{Bull}_{t-2} = 1$$
$$\text{Bull3}_t = \text{Bull3}_{t-1} \text{ otherwise.}$$

TABLE 9.2

Monthly Excess RORs, cont'd

	Mo.	Market	Fund	Bull	S3	Bull3
3	J	−1.13%	−1.80%	0	1	0
yrs	A	1.07%	1.52%	1	1	0
ago,	S	−9.61%	−9.98%	0	1	0
cont'd	O	−18.62%	−18.24%	0	1	0
	N	−7.80%	−8.80%	0	0	0
	D	0.78%	1.35%	1	1	0
2	J	−8.97%	−9.69%	0	1	0
yrs	F	−11.67%	−9.11%	0	1	0
ago	M	8.18%	7.58%	1	1	0
	A	8.96%	8.61%	1	2	0
	M	5.16%	5.51%	1	3	1
	J	0.00%	0.31%	1	3	1
	J	7.14%	8.69%	1	3	1
	A	3.29%	3.64%	1	3	1
	S	3.50%	4.34%	1	3	1
	O	−2.00%	−2.37%	0	2	1
	N	5.57%	4.67%	1	2	1
	D	1.76%	2.91%	1	2	1
1	J	−3.77%	−4.13%	0	2	1
yr	F	2.80%	4.51%	1	2	1
ago	M	5.70%	6.28%	1	2	1
	A	1.45%	1.32%	1	3	1
	M	−8.57%	−8.55%	0	2	1
	J	−5.55%	−5.81%	0	1	1
	J	6.64%	7.09%	1	1	1
	A	−4.87%	−5.05%	0	1	1
	S	8.38%	8.96%	1	2	1
	O	3.61%	2.45%	1	2	1
	N	−0.24%	0.59%	0	2	1
	D	6.31%	7.31%	1	2	1

In programming this, it is convenient to work in terms of the moving sum

$$S3_t = \text{Bull}_t + \text{Bull}_{t-1} + \text{Bull}_{t-2}.$$

Then

$$\text{Bull3}_t = 0 \text{ if } S3_t = 0,$$
$$\text{Bull3}_t = 1 \text{ if } S3_t = 3, \text{ and}$$
$$\text{Bull3}_t = \text{Bull3}_{t-1} \text{ if } S3_t = 1 \text{ or } 2.$$

Table 9.1 shows the RORs and the Bull3 indicator. Table 9.2 is a continuation of this table.

9.2.2 Regressions on Bull3

Next some regressions on Bull3 are shown, namely:

- **Two betas, no alpha:** Regression of fund on market and market*Bull3 (no constant)

- **Two betas, one alpha:** Regression of fund on market and market*Bull3 (with constant)

- **Two betas, two alphas:** Regression of fund on market, Bull3 and market*Bull3 (with constant)

9.2.2.1 Two Betas, No Alpha

This is the regression of Y = excess ROR of the fund on X_1 = excess ROR of the market and $X_2 = X_1 \times$ Bull3 (without a constant).

```
Regression Analysis: fund versus market, market*Bull3

The regression equation is
fitted value of fund = 0.975 market + 0.0939 market*Bull3


Predictor          Coef   SE Coef      t      p
Noconstant
market          0.97463   0.02259  43.15  0.000
market*Bull3    0.09388   0.03700   2.54  0.014

s = 0.724429

Analysis of Variance
Source            DF       SS       MS       F      p
Regression         2  1674.80   837.40  1595.66  0.000
Residual Error    58    30.44     0.52
Total             60  1705.24
```

The fitted regression $\hat{Y} = b_1 x_1 + b_2 x_2$ is

fitted value of fund = 0.975 market + 0.0939 market*Bull3.

This is

$$
\begin{aligned}
\text{fitted value of fund} &= 0.975 \text{ market, if Bull3} = 0 \\
&= 1.069 \text{ market, if Bull3} = 1.
\end{aligned}
$$

The standard error of fit is 0.724.

9.2.2.2   Two Betas, One Alpha

This is the regression of Y = excess ROR of the fund on X_1 = excess ROR of the market and $X_2 = X_1 \times$ Bull3 (with constant). The fitted regression will be of the form $\hat{Y} = a + b_1x_1 + b_2x_2$.

```
Regression Analysis: fund versus market, market*Bull3

The regression equation is
fitted value of fund = 0.171 + 0.983 market + 0.0729 market*Bull3

Predictor        Coef  SE Coef      t      p
Constant      0.17081  0.09671   1.77  0.083
market        0.98350  0.02275  43.24  0.000
market*Bull3  0.07286  0.03824   1.91  0.062

S = 0.711544   R-Sq = 98.3%   R-Sq(adj) = 98.2%

Analysis of Variance
Source          DF      SS      MS       F      p
Regression       2  1676.16  838.08  1655.32  0.000
Residual Error  57    28.86    0.51
Total           59  1705.02
```

The fitted regression $\hat{Y} = a + b_1x_1 + b_2x_2$ is

fitted value of fund = 0.171 + 0.983 market + 0.0729 market*Bull3.

This is

$$
\begin{aligned}
\text{fitted value of fund} &= 0.173 + 0.983 \text{ market, if Bull3} = 0 \\
&= 0.173 + 1.056 \text{ market, if Bull3} = 1.
\end{aligned}
$$

The standard error of fit is 0.7115.

9.2.2.3   Two Betas, Two Alphas

This is the regression of Y = excess ROR of the fund on X_1 = Bull3, X_2 = excess ROR of the market and $X_3 = X_2 \times$ Bull3 (with constant), that is, the regression of fund on market, Bull3, and market*Bull3 (with constant):

```
Regression Analysis: fund versus market, Bull3, market*Bull3

The regression equation is
fitted value of fund = 0.002 + 0.975 market + 0.224 Bull3 + 0.0778 market*Bull3
```

```
Predictor          Coef  SE Coef       t     p
Constant         0.0018   0.1957    0.01  0.993
market          0.97472  0.02440   39.94  0.000
Bull3            0.2236   0.2251     0.99  0.325
market*Bull3    0.07775  0.03856    2.02  0.049


S = 0.711626   R-Sq = 98.3%   R-Sq(adj) = 98.2%

Analysis of Variance
Source           DF       SS       MS        F      p
Regression        3  1676.66   558.89  1103.62  0.000
Residual Error   56    28.36     0.51
Total            59  1705.02
```

The fitted model is

fitted value of fund $= 0.002 + 0.9747$ market $+ 0.224$Bull3 $+ 0.0778$ market*Bull3.

If Bull3 $= 0$, this is

$$\text{fitted value of fund } = 0.002 + 0.9747 \text{ market}.$$

If Bull3 $= 1$, this is

$$
\begin{aligned}
\text{fitted value of fund} &= (.002 + .224) + (0.9747 + 0.0778) \text{ market} \\
&= .226 + 1.0525 \text{ market}.
\end{aligned}
$$

The standard error of fit is 0.7116, about the same as the 0.7115 obtained with the preceding model.

9.2.3 Other Models for Bull/Bear

9.2.3.1 Two Means and Two Variances

A model with separate Bull and Bear distributions can be considered. There can be a mean and variance for the Bull state and another mean and variance for the Bear state. The parameters of the two distributions can be estimated from the months with Bull3 $= 0$ and those with Bull3 $= 1$. The results are shown below. The mean and standard deviation for Bear months are -3.34% and 7.53%, respectively; those for Bull months are $+0.988\%$ and 3.635%, respectively. Thus as expected, there is a negative mean and relatively large standard deviation for Bear months and a positive mean and relatively small standard deviation for Bull months. The size of the coefficient of variation–ratio of standard deviation to absolute value of the mean–is $7.53/3.34 = 2.25$ for Bear months and $3.635/0.988 = 3.68$ for Bull months.

```
Descriptive Statistics: market
```

Bull3	N	N*	Mean	SE Mean	StDev	Min	Q1	Median	Q3	Max
0	16	0	-3.34	1.88	7.53	-18.62	-9.10	-2.42	1.03	8.96
1	44	0	0.988	0.548	3.635	-8.570	-0.938	1.065	3.265	8.38

The preceding is predicated upon a given indicator for Bull/Bear states. Indicators such as Bull and Bull3 seem somewhat arbitrary. What about a Bull4, based on a moving window of width 4, or a Bull2, based on a moving window of width 2? Why not use a more flexible method that lets the data do more to speak for themselves? To do this, we consider other descriptions, which model Bull and Bear as hidden (latent) states, to be detected from the data.

9.2.3.2 Mixture Model

Another model that incorporates hidden (latent) states and fits state-specific distributions is the *finite mixture model*. (See, for example, McLachlan and Peel 2000). This model is for any number K of states and observations $\boldsymbol{x}_t$ that may be vectors. The p.d.f. is of the form

$$
\begin{aligned}
f(\boldsymbol{x}_t) \quad = \quad & \pi_1\, f_1(\boldsymbol{x}_t) + \pi_2\, f_2(\boldsymbol{x}_t) + \cdots \\
& + \pi_k\, f_k(\boldsymbol{x}_t) \; + \; \cdots \; + \; \pi_K\, f_K(\boldsymbol{x}_t), \; t = 1, 2 \ldots, n,
\end{aligned}
$$

where the observations are at n time points. The p.d.f.s f_k are called *component* p.d.f.s or *state-conditional* p.d.f.s. Sometimes the states are called *classes* and the corresponding distributions are called *class-conditional distributions*. The *mixture probabilities* π_k add to 1. The mixture probability π_k is the prior probability that an observation comes from class k.

When there are just two states, it is often convenient to index them by $k = 0, 1$. The p.d.f. is

$$
f(\boldsymbol{x}_t) \; = \; \pi_0\, f_0(\boldsymbol{x}_t) + \pi_1 f_1(\boldsymbol{x}_t),
$$

where the mixture probabilities π_0 and π_1 add to 1. The two class-conditional distributions could be multivariate Normal. Of course, the vector $\boldsymbol{x}$ can be just a scalar x. The class-conditional densities could, for example, be Normal, with $f_k(x) = \phi(x; \mu_k, \sigma_k^2)$, $k = 0, 1$, where $\phi(x; \mu, \sigma^2)$ is the p.d.f. of the Normal distribution with mean μ and variance σ^2. The probabilities $\pi_k, k = 0, 1$ are prior probabilities of the distributions given by f_0 and f_1. The posterior probability of state k, given x, is

$$
\pi(k \,|\, x) \; = \; \frac{\pi_k\, f_k(x)}{f(x)} \; = \; \frac{\pi_k f_k(x)}{\pi_0 f_0(x) + \pi_1 f_1(x)}, \; k = 0, 1.
$$

In the Gaussian case this is

$$
\pi(k \,|\, x) \; = \; \frac{\pi_k\, \phi(x; \mu_k, \sigma_k^2)}{\pi_0 \phi(x; \mu_0, \sigma_0^2) + \pi_1 \phi(x; \mu_1, \sigma_1^2)}, \; k = 0, 1.
$$

Denote estimates of $\pi_k, \mu_k, \sigma_k^2, \pi(k\,|\,x)$ by $p_k, m_k, v_k, p(k\,|\,x)$. Estimation starts with initial guesses $p_k^{(0)}, m_k^{(0)}, v_k^{(0)}, k = 0, 1$. Then it alternates between estimating the posterior probabilities and the distributional parameters. (This alternation is an example of the EM algorithm; see Dempster, Laird and Rubin 1978; McLachlan and Krishnan 1997). At stage N ($N = 0, 1, 2, \ldots$) set

$$p^{(N+1)}(k|x) = \frac{p_k^{(N)}\phi(x; m_k^{(N)}, v_k^{(N)})}{p_0^{(N)}\phi(x; m_0^{(N)}, v_0^{(N)}) + p_1^{(N)}\phi(x; m_1^{(N)}, v_1^{(N)})}, \quad k = 0, 1,$$

$$p_k^{(N+1)} = \sum_{t=1}^{n} p^{(N+1)}(k\,|\,x_t)/n, \quad k = 0, 1,$$

$$m_k^{(N+1)} = \sum_{t=1}^{n} w^{(N+1)}(k|x_t)\, x_t, \quad v_k^{(N+1)} = \sum_{t=1}^{n} w^{(N+1)}(k|x_t)\, (x_t - m_k^{(N)})^2,$$

where the weights $w^{(N+1)}(k\,|\,x_t)$ are proportional to the estimated posterior probabilities,

$$w^{(N+1)}(k\,|\,x_t) = p^{(N+1)}(k\,|\,x_t)/\sum_{t=1}^{n} p^{(N+1)}(k\,|\,x_t).$$

Iterations are continued until adequate precision is reached.

9.2.3.3 Hidden Markov Model

Another model with hidden (latent) states is the *hidden Markov model* (HMM). Like a finite mixture model, an HMM includes state-conditional probability functions, for example, Normal distributions with different means and different variances, but in addition there is a matrix of *transition probabilities*. The entry π_{jk} of the matrix gives the probability of transition to state k, given that the process was in state j in the preceding time period,

$$\pi_{jk} = \Pr\{S_t = k\,|\,S_{t-1} = j\}, \; j, k = 1, 2, \ldots, \; K \text{ states},$$

where $\{S_t\}$ is the unobservable state process. Each $S_t = 1, 2, \ldots,$ or K.

The p.d.f. for an HMM is analogous to that of a mixture model, but with transition probabilities taking the place of mixture probabilities:

$$f(x_t\,|\,S_{t-1} = j) = \pi_{j1}f_1(x_t) + \pi_{j2}f_2(x_t) + \cdots + \pi_{jk}f_k(x_t) + \cdots + \pi_{jK}\,f_K(x_t).$$

For modeling Bull and Bear states, we take $k = 0, 1$ and use Normal p.d.f.s, obtaining the p.d.f.

$$f(x_t\,|\,S_{t-1} = k) = \pi_{k0}\,\phi(x_t; \mu_0, \sigma_0^2) + \pi_{k1}\,\phi(x_t; \mu_1, \sigma_1^2), \; k = 0, 1,$$

where S_t denotes the hidden (unobservable) state, 0 for Bear, 1 for Bull, at time t.

Remarks. (i) Here, an HMM is being applied to RORs. It is often noted that the distribution of RORs is not Gaussian. Note, however, that although the marginal distribution of x_t may not be Gaussian, the component densities $f_k(x_t)$ in a finite mixture model or HMM may be. (ii) In the above formulation, the transition probabilities are taken as *stationary* (not time dependent, that is, the corresponding stochastic process is *homogeneous*), but they may instead be modeled in terms of time-varying covariates, such as the risk-free rate.

HMMs for Bull and Bear markets. Maheu and McCurdy (2000) used a Markov-switching model that incorporates duration dependence to capture nonlinear structure in both the conditional mean and the conditional variance of stock returns. Their data are monthly returns including dividends (1802 to 1925 from Schwert 1990; 1926 to 1995:12 from CRSP, the Center for Research in Security Prices). The model sorts returns into a high-return stable state and a low-return volatile state. These can be labeled as Bull and Bear states, respectively. The authors' method identifies all major stock-market downturns in over 160 years of monthly data. Bull markets have a declining hazard function although the best market gains come at the start of a Bull market. Volatility increases with duration in Bear markets. Allowing volatility to vary with duration captures volatility clustering. According to their method, the market spent 90% of the time as a Bull market and only 10% in a Bear market.

Lunde and Timmermann (2004) studied time series dependence in the direction of stock prices by modeling the (instantaneous) probability that a Bull or Bear market terminates as a function of its age and a set of underlying state variables, such as interest rates. A random walk model is rejected for both Bull and Bear markets. Although it fits the data better, a generalized autoregressive conditional heteroscedasticity model (GARCH) is also found to be inconsistent with the very long Bull markets observed in the data. The strongest effect of increasing interest rates is found to be a lower Bear market hazard rate (chance of ending, as time goes on) and hence a higher likelihood of continued declines in stock prices.

Huang and Sclove (2011) segmented the time series of S&P500 monthly RORs using an HMM with Gaussian p.d.f.s with differing means and variances. Three values of the number K of states were tried: $K = 1, 2, 3$. With K states, the number of free parameters is $2K$ means and variances and $K(K-1)$ free parameters in the transition probability matrix, a total of $2K + K(K-1) = K(K-1+2) = K(K+1)$. This is 6 for $K = 2$, 12 for $K = 3$, and just 2 for $K = 1$. A conclusion that $K = 1$ fits best corresponds to no state-switching and might justify fitting a single ARIMA model over all time points. A conclusion that $K = 2$ corresponds to the conventional use of two states, Bull and Bear, particularly if one state has a positive mean ROR and the other has a negative mean. A conclusion that $K = 3$ would suggest another state, perhaps intermediate between Bull and Bear. BIC (Schwarz 1978) was used to compare models.

BIC_k is an expansion of $-2 \ln(\text{posterior probability of Model } k)$, and so can be put onto a scale of posterior probability. That is, the posterior probability of Model k, $p(k \,|\, \text{data})$, is approximately proportional to $\exp(-\text{BIC}_k/2)$. (This assumes equal prior probabilities of $1/K$ but the expression can be adjusted for unequal prior probabilities.) That is,

$$p(k \,|\, \text{data}) \approx \exp[-\text{BIC}_k/2] \,/ \sum_{j=1}^{K} \exp[-\text{BIC}_j/2].$$

It was found that the conventional number of states $K = 2$ has the best (lowest) BIC and hence the highest posterior probability, with a larger variance (and smaller mean) corresponding to the Bear state.

Formulation of HMMs. Next the likelihood function for HMMs will be discussed.

Joint distribution. The joint distribution of the observations in a time series may be built up from

$$f(\boldsymbol{x}_1, \boldsymbol{x}_2, \ldots, \boldsymbol{x}_n) = \\ f(\boldsymbol{x}_1) f(\boldsymbol{x}_2 \,|\, \boldsymbol{x}_1) f(\boldsymbol{x}_3 \,|\, \boldsymbol{x}_2, \boldsymbol{x}_1) \cdots f(\boldsymbol{x}_n \,|\, \boldsymbol{x}_n, \boldsymbol{x}_{n-1}, \ldots, \boldsymbol{x}_2, \boldsymbol{x}_1).$$

In the case of a first order Markov process, this becomes

$$f(\boldsymbol{x}_1, \boldsymbol{x}_2, \ldots, \boldsymbol{x}_n) = f(\boldsymbol{x}_1) f(\boldsymbol{x}_2 \,|\, \boldsymbol{x}_1) f(\boldsymbol{x}_3 \,|\, \boldsymbol{x}_2) \cdots f(\boldsymbol{x}_n \,|\, \boldsymbol{x}_n).$$

The joint distribution for an HMM takes account of both the observations and the hidden states. The observation vector $\boldsymbol{x}_t$ is $(y_t \ s_t)$. Thus, $f(\boldsymbol{x}_t \,|\, \boldsymbol{x}_{t-1}) = f(y_t, s_t \,|\, y_{t-1}, s_{t-1}) = f(y_t \,|\, s_t, y_{t-1}, s_{t-1}) f(s_t \,|\, y_{t-1}, s_{t-1}) = f(y_t \,|\, s_t) f(s_t \,|\, s_{t-1}) = f_{s_t}(y_t) \pi_{s_{t-1}, s_t}$. Here we have used the facts of the HMM model formulation that $f(y_t \,|\, s_t, y_{t-1}, s_{t-1}) = f(y_t \,|\, s_t) = f_{s_t}(y_t)$ and $f(s_t \,|\, y_{t-1}, s_{t-1}) = f(s_t \,|\, s_{t-1}) = \pi_{s_{t-1}, s_t}$. For a given state sequence the probability is

$$p(s_0) \, \pi_{s_0, s_1} \, f(y_1 \,|\, s_1) \, \pi_{s_1 s_2} f(y_2 \,|\, s_2) \cdots \pi_{s_{n-1} s_n} f(y_n \,|\, s_n).$$

This involves also an initial distribution $p(s_0), s_0 = 1, 2, \ldots, K$. The marginal distribution of the observable $y_1, y_2, \ldots, y_n$ is a sum of this over all such possible state sequences, because the event that $y_1, y_2, \ldots, y_n$ occurs is the union of their occurrence over possible state sequences.

The Likelihood. The likelihood is the joint distribution of the observations, with the observations taken as given and the parameters taken as variables. The likelihood is to be maximized over the parameters, both the distributional parameters and the transition probabilities.

Estimating the Parameters of an HMM.

The parameters may be estimated by direct maximization of the likelihood or by the *Baum/Welch algorithm* (Baum and Welch (ca. 1970); Baum, Petrie,

Soules, and Weiss 1970). In the Baum, Petrie, Soules, and Weiss article (1970), there is a reference to a paper by Baum and Welch submitted to the *Proceedings of the National Academy of Sciences* (PNAS), but the archives of PNAS reveal no publication by Baum and Welch. The Baum/Welch algorithm has become an important tool in many fields, first especially in speech recognition. The algorithm estimates the probability distribution $\pi_t(k) = \Pr\{S_t = k\}$ over the states of the hidden Markov chain at each time point. Though formulated before any general formulation of the EM algorithm, the Baum/Welch algorithm is an EM algorithm. We do not elaborate further on the Baum/Welch algorithm here. See, for example, McLachlan and Krishnan (1997).

9.2.4   Bull and Bear Portfolios

Having segmented the relevant time series of market indicators into states such as Bull and Bear, one may then estimate the mean vector and covariance matrix of a portfolio of selected stocks and form a portfolio that is optimal in Bull markets and a portfolio that is optimal in Bear markets, using formulas such as those in earlier chapters on portfolio optimization. The estimate of an optimal weight vector w is a function of the estimate m of the mean vector and S of the covariance matrix, say $\hat{w}^* = H(m, S)$. There are estimates m_0 and S_0 and m_1, S for Bull and Bear states, respectively; $\hat{w}_0^* = H(m_0, S_0)$, $\hat{w}_1^* = H(m_1, S_1)$. One would try to predict whether or not there will be a change in the state of the market, and rebalance or not, accordingly. A decision risk analysis should be performed to see if the expected increase in portfolio ROR would compensate for transaction costs incurred in rebalancing. The transaction costs might be proportional to the amount of buying and selling across the m stocks, that is, proportional to $\sum_{a=1}^{m} |\hat{w}_{1a}^* - \hat{w}_{0a}^*|$, where w_{ka} is the weight of asset a in the portfolio for state k, $k = 0, 1$.

Example 9.1 /rm

GDP: Gross Domestic Product

   Next, a macro-economic example, GDP, will be considered further. GDP was considered in the preceding chapter, on time-series analysis, where a seasonal ARIMA model for log quarterly GDP was discussed. It is interesting to compare this with a model with different states, although even if it happened that a single ARIMA model fit better, one might still want to fit a model with states, which labels the time points.

   Chen and Sclove (2011) fit a state model to GDP. The time series of quarterly growth rates of US GDP, from 1947 through the third quarter of 2010, was segmented by hidden Markov models (HMMs). HMMs with several states were fit and compared with a single distribution for the growth rate. State-conditional Normal distributions with different means and variances were fit for different numbers of states. The extent to which states correspond to recession, recovery, expansion, and contraction was assessed. The HMMs were scored by BIC (Schwarz 1978). Some comparison was made to ARIMA models involving regular and quarterly differencing and regular and quarterly autoregression of log GDP. Components of GDP were also fit with HMMs, with a view toward determining which components are leading or lagging indicators of the state of overall GDP.

9.3 Summary

Time series may have *states*. These are phases through which the process passes over time. Examples are Bull and Bear states of the stock market, and Recession, Recovery, Expansion, and Contraction in the economy.

The model is that at each time point there is an observable variable and a state variable, the latter indicating the phase of the process at that time point. The states may be unobservable but inferred from the observed values. A *segment* (*regime, epoch*) is a sequence of time points having the same state. *Segmentation* involves estimating the states, that is, labeling each time point with its most likely state, or giving a probability distribution over the states for each time point.

Hidden Markov models (HMMs) involve state-conditional p.d.f.s and transitions between states. The state process is a Markov chain with a *transition probability matrix*.

Alternative, competing models can be compared with criteria such as BIC (Bayesian Information Criterion). The values of BIC for the alternative models can be put on a scale of posterior probability.

9.4 Chapter Exercises

9.4.1 Applied Exercises

9.1 From finance.yahoo.com, or otherwise, obtain data for the most recent five years on the Dreyfus fund with ticker symbol DREVX and the S&P500 (ticker symbol ^GSPC; this is "hat" followed by GSPC). Obtain the rates for three-month Treasury bills (TB3MS). Compute the excess RORs. Do the regression of the fund on the Bull indicator. Compare the results with those obtained in Chapter 5.

9.2 (continuation) Compute the mean and standard deviation for the Bull months and for the Bear months. Which has the larger standard deviation?

9.3 Obtain data for the most recent five years on the Dreyfus fund DREVX and the S&P500 ^GSPC. Obtain the rates for three-month Treasury bills (TB3MS). Compute the excess RORs. Do the regression of the fund on the Bull3 indicator. Compare the results with those obtained in this chapter.

9.4 (continuation) Compute the mean and standard deviation for the Bull months and for the Bear months. Which has the larger standard deviation?

9.5 Repeat the preceding two exercises for another mutual fund.

9.6 Choose two stocks. Compute their Bull and Bear betas by any of the methods discussed. Form Sharpe-ratio optimal Bull and Bear portfolios. (This is probably most interesting if the RORs of the two stocks are negatively correlated, or at least not highly positively correlated.)

9.7 Choose three stocks. Compute their Bull and Bear betas by any of the methods discussed. Form Sharpe-ratio optimal Bull and Bear portfolios. (This is probably most interesting if there is one negative correlation among the three pairwise correlations of the RORs of the stocks.)

9.4.2 Mathematical Exercises

9.8 Infinite mixture. We use the term *finite* mixture model because there is such a thing as an infinite mixture model. As an example, show that the p.d.f. of Student's t is a scale mixture of Gaussian p.d.f.s, with the reciprocal variance of the Gaussian family taken as having a Gamma distribution. (See Section 2.4.1.)

9.9 Compound distribution. Show that if the Poisson parameter λ has a Gamma distribution, the resulting marginal distribution is Negative Binomial. This sort of *compound distribution* is another example of an infinite mixture. (See Section 2.4.2.)

9.10 Infinite mixture of Poissons. Interpret the result of the preceding problem in terms of the number of claims of insureds having different accident rates (accident proneness) with a Gamma distribution over their individual accident rates. (See Section 2.4.2.)

9.5 Bibliography

Baum, Leonard E., and Welch, Lloyd. A statistical estimation procedure for probabilistic functions of finite Markov processes. Submitted ca. 1970 for publication to *Proceedings of the National Academy of Sciences of the U.S.A.*

Baum, Leonard E., Petrie, Ted, Soules, George, and Weiss, Norman (1970). A maximization technique occurring in the statistical analysis of probabilistic functions of Markov chains. *Annals of Mathematical Statistics*, **41**, 164–171.

Chen, Yu, and Sclove, Stanley L. (2011). Segmenting the time series of quarterly GDP using a hidden Markov model. Presentation at the *Joint Statistical Meetings* (American Statistical Association and related societies), Miami Beach, FL.

Dempster, A. P., Laird, N. M., and Rubin, D. B. (1977). Maximum likelihood from incomplete data via the EM algorithm. *Journal of the Royal Statistical Society, Series B (Methodological)*, **39**, 1–38.

Forney, G. D. (1973). The Viterbi algorithm. *Proceedings of the IEEE*, **61**, 268–278.

Huang, Ziqian, and Sclove, Stanley L. (2011). Segmenting the time series of a market index using a hidden Markov model. Submitted to *Journal of Classification*.

Investopedia (2011). Bear market. www.investopedia.com/terms/b/bearmarket.asp#axzz1gXPbjeE2

Kashyap, Rangasami L. (1982). Optimal choice of AR and MA parts in autoregressive moving average models. *IEEE Transactions on Pattern Analysis and Machine Intelligence*, **4**, 99–104.

Lunde, Asger, and Timmermann, Allan (2004). Duration dependence in stock prices: An analysis of Bull and Bear markets. *Journal of Business and Economic Statistics*, **22**, 253–273.

Maheu, John M., and McCurdy, Thomas H. (2000). Identifying Bull and Bear markets in stock returns. *Journal of Business and Economic Statistics*, **18**, 100–112.

McLachlan, Geoffrey, and Krishnan, Thiryambakam (1997). *The EM Algorithm and Extensions*. John Wiley & Sons, New York.

McLachlan, Geoffrey, and Peel, David (2000). *Finite Mixture Models*. John Wiley & Sons, New York.

Schwarz, Gideon (1978). Estimating the dimension of a model. *Annals of Statistics,* **6,** 461–464.

Schwert, G. W. (1990). Indexes of U.S. stock prices from 1802 to 1987. *Journal of Business,* **63,** 399–426.

Vangard Group (2011). Staying calm during a Bear market. retirementplans.vanguard.com/VGApp/pe /PubVgiNews?ArticleName=Stayingcalmbearmkt.

Viterbi, A. J. (1967). Error bounds for convolutional codes and an asymptotically optimum decoding algorithm. *IEEE Transactions on Information Theory,* **13,** 260–269.

Zucchini, Walter, and MacDonald, Iain L. (2009). *Hidden Markov Models for Time Series: an Introduction using R.* Chapman & Hall/CRC Press/Taylor & Francis Group, Boca Raton, FL.

9.6 Further Reading

The model selection criterion BIC was formulated by Schwarz (1978); the paper by Kashyap (1982) is also very helpful and gives more details.

To begin a study of HMMs, it is helpful to begin with the Viterbi algorithm, which finds the most likely state sequence. See Viterbi (1968) and Forney (1978). Then for further study on the HMM, one may refer, for example, to McLachlan and Krishnan (1997) or Zucchini and MacDonald (2009).

A

Vectors and Matrices

CONTENTS

This appendix is mainly a listing of background results that can be used for deeper understanding of the text material. There is a brief discussion and proof at times. Some of the results discussed here were presented also earlier in the text where needed.

A.1   Introduction

This appendix deals with *arrays*. A two-dimensional array is a *matrix*. An $m \times n$ (read "m-by-n") matrix may be written as having m rows and n columns, the rows being horizontal lines, the columns being vertical lines.

Each line of a matrix is a *vector*. An $m \times 1$ array is a *column* vector. A $1 \times n$ array is a *row* vector.

A vector $\boldsymbol{v}$ with elements $v_1, v_2, \ldots, v_m$, say, may be pictured as a one-dimensional array, but its essence is that it is a function from the set of subscripts i ($i = 1, 2, \ldots, m$, say) to the objects (usually numbers) v_i that are the elements of the vector. Similarly, a matrix $\boldsymbol{A}$ with elements a_{ij}, $i = 1, 2, \ldots, m$, $j = 1, 2, \ldots, n$, may be pictured as a two-dimensional array, but its essence is that it is a function from the set of pairs (i, j) to the numbers $a(i, j)$ that are the elements of the matrix.

A.2   Vectors

A *vector*, then, is an ordered set of elements (usually numbers in our case).

For example, if John buys three items at the grocery store, two dozen eggs at \$1.00 per dozen, three loaves of bread at \$1.50 per loaf, and four cans of soup at \$0.60 per can, then the prices and quantities can be taken as vectors

$$\boldsymbol{p} = (p_1\ p_2\ p_3) = (\$1.00\ \$1.50\ \$0.60),$$

and

$$\boldsymbol{q} = (q_1\ q_2\ q_3) = (2\ 3\ 4),$$

where the subscript 1 denotes eggs, 2 denotes bread, and 3 denotes soup.

A.2.1   Inner Product of Two Vectors

At the cash register, John will owe

$$\$1.00 \times 2 + \$1.50 \times 3 + \$0.60 \times 4 = \$2.00 + \$4.50 + \$2.40 = \$8.90.$$

This operation is a sum of products of corresponding elements of the price and quantity vectors.

The *inner product* or *dot product,* or *scalar product* $\boldsymbol{u} \cdot \boldsymbol{v}$ of two vectors

$u = (u_1 \ u_2 \ ... \ u_n)$ and $v = (v_1 \ v_2 ... \ v_n)$ of the same dimension, n, is the sum of products of corresponding elements:

$$u \cdot v = u_1 v_1 + u_2 v_2 + ... + u_n v_n.$$

Above we had $n = 3$, $u = p$, and $v = q$; the inner product $u \cdot v = p \cdot q$ is the total bill.

Remark. We often write $u \cdot v$ as $u'v$. This is often considered a row vector times a column vector, although all that is essential is that it is a sum of products of corresponding elements.

A.2.2 Orthogonal Vectors

To plot a vector, draw an arrow from the origin to the point indicated by the vector's coordinates. Plot the vectors (1 1) and (1 − 1). You will note that these two vectors are *orthogonal* (perpendicular). The angle between them is a right angle.

Two vectors $u = (u_1 \ u_2 \ ... \ u_p)$ and $v = (v_1 \ v_2 ... \ v_p)$ of the same dimension are *orthogonal* if their dot product is 0.

The "arrows" representing two such vectors are at right angles to one another.

A.2.3 Variates

In many fields, including statistics, economics, and finance, *linear combinations* of variables play a major role.

Given variables $v_1, v_2, ..., v_p$, a *linear form* in those variables is a linear combination of those variables, that is, a function of the form

$$f(v_1, v_2, \ldots, v_p) = a_1 v_1 + a_2 v_2 + \ldots + a_p v_p,$$

where $a_1, a_2, \ldots, a_p$ are constants. In statistics this linear combination is sometimes called a *variate*. It can be written in terms of vectors as

$$a'v.$$

A special case is a *weighted average,* where

$$a_i = w_i > 0 \text{ and } \sum_{i=1}^{n} w_i = 1.$$

Sometimes a constant is added: $a_0 + a'v$.

A.2.4 Section Exercises

A.1 On graph paper draw the vectors $(1, 1)$ and $(1, -1)$. Note that their inner product is zero. Notice on the graph paper that the angle between these two vectors is a right angle.

A.2 Find two vectors orthogonal to $(1, 1, 1)$.

A.3 Find two vectors orthogonal to $(1, 1, 1)$ and to each other.

A.4 Find two vectors orthogonal to $(1, 2, 3)$.

A.5 Find two vectors orthogonal to $(1, 2, 3)$ and to each other.

A.6 Show that $a_0 + \boldsymbol{a}'\boldsymbol{v} = (a_0\ \boldsymbol{a})'(1\ \boldsymbol{v})$.

A.7 Show that $(1\ 1\ 1)'(v_1\ v_2\ v_3) = \sum_{i=1}^{3} v_i$. That is, the inner product of a vector with the vector of ones equals the sum of the elements of the vector.

A.8 Show that $(1\ 0\ 0)'(v_1\ v_2\ v_3) = v_1$. That is, the inner product of a vector with the vector with its first element 1 and the other elements 0 picks out the first element of the vector. (Similarly, the inner product of a vector and the vector with its second element 1 and the other elements 0 picks out the vector's second element.)

A.9 (In regard to the next exercises, see also Section A.5 on Vector Differentiation below.) What is the partial derivative of $\boldsymbol{a}'\boldsymbol{v}$ with respect to v_1?

A.10 What is the partial derivative of $\boldsymbol{a}'\boldsymbol{v}$ with respect to v_2?

A.11 The notation $\mathbf{grad}\, f$ (*gradient* of f) or $\frac{\partial f(\boldsymbol{v})}{\partial \boldsymbol{v}}$ denotes the p-vector of partial derivatives of f with respect to each of the entries of $\boldsymbol{v}$. What is this if f is the variate $\boldsymbol{a}'\boldsymbol{v}$?

A.3   Matrices

A *matrix* is a rectangular array of numbers. The *lines* of a matrix are called *rows* and *columns*. A matrix can be considered a set of column vectors or a set of row vectors, that is, the *lines* of a matrix are called *rows* and *columns*.

Vectors are matrices of one row or one column.

It is convenient to denote matrices by uppercase boldface, for example, $\boldsymbol{M}$, and vectors by lowercase boldface, for example, $\boldsymbol{v}$.

A.3.1 Entries of a Matrix

If M is an $I \times J$ matrix with elements (or *entries*) $m_{ij}, i = 1, 2, \ldots, I$, and $j = 1, 2, \ldots, J$, then we write

$$M = [m_{ij}]_{i=1,2,\ldots,I; j=1,2,\ldots,J}.$$

A.3.2 Transpose of a Matrix

First recall that if v is a vertical (column) vector, then the transpose v' denotes the same elements arranged as a horizontal (row) vector.

If M is a matrix with elements $m_{ij}, i = 1, 2, ..., I$, and $j = 1, 2, ..., J$, then its *transpose* is the $J \times I$ matrix

$$M' = [m_{ji}]_{j=1,2,\ldots,J; i=1,2,\ldots,I}.$$

(For vectors, usually v will be taken to be a column vector; v ', a row vector.)

The columns of M become the rows of M'.

A.3.3 Matrix Multiplication

If A is $m \times n$ and B is $n \times p$, then their product AB is the $m \times p$ matrix whose (i, j)-th entry is the inner product of the i-th row of A and the j-th column of B, for $i = 1, 2, \ldots, m;\ j = 1, 2, \ldots, p$.

Note. Here the word "matrix" is used as the adjectival form of the noun *matrix*, as in the phrase *matrix algebra*. Sometimes the adjectival form is taken as "matric," as in *matric algebra*.

In addition to matrix addition, subtraction, and multiplication, an additional operation is multiplication of a matrix by a scalar. In the context of vectors and matrices, a number is called a *scalar*. Let $A = [a_{ij}]$ be an m-by-n matrix and let c be a number. Then the matrix $c \cdot A$, or simply cA, is the $m \times n$ matrix obtained by multiplying each entry of A by c:

$$c A = [c\,a_{ij}].$$

The matrix cA is called a *scalar multiple* of A.

A.3.4 Section Exercises

A.12 Peter and Paul go to the grocery store. They each buy the same three items: eggs, bread, and canned soup. Peter buys two dozen eggs, one loaf of bread, and five cans of soup. Paul buys one dozen eggs, two loaves of bread, and three cans of soup. The prices are \$1.00 per dozen eggs, \$1.50 per loaf of bread, and \$0.60 per can of soup. Let B be the 3×1 vector of prices and A be the 2×3 matrix of quantities, two rows for the boys, and three columns for the items. Compute and interpret the matrix product $A\,B$.

A.13 (continuation) Recompute the matrix product if the prices of eggs is $0.50 per dozen.

A.3.5   Identity Matrix

The *identity matrix* of order n, denoted by I, is the $n \times n$ matrix with 1's on the main diagonal and 0's elsewhere. Its products are given by $AI = A$ and $IB = B$.

A.3.6   Inverse

A.3.6.1   Inverse of a Matrix

Given a square matrix A, if there is a matrix B such that $AB = I$, then the matrix B is called the *inverse* of A. Then $BA = I$ as well.

The inverse of the matrix M is denoted by M^{-1}. So the preceding statement says that

$$A^{-1}A = I = AA^{-1}.$$

A.3.6.2   Inverse of a Product of Matrices

First of all, suppose you put on your shirt, and then you put on your sweater over that. To reverse this, you must first take off the sweater, then the shirt.

To see this for matrices and linear transformations, let A be $m \times n$ and B be $n \times p$. Let $C = AB$. Now suppose $y = Cx$. We shall see that

$$C^{-1} = B^{-1}A^{-1},$$

that is, the inverse of a product is the product of the inverses, in reverse order.

Suppose that $y = Cx$; that is, $y = ABx$. Then $x = B^{-1}A^{-1}y$. To see this, pre-multiply $y = ABx$ by A^{-1}, obtaining

$$A^{-1}y = A^{-1}ABx = IBx = Bx.$$

Now, pre-multiply by B^{-1}, obtaining

$$B^{-1}A^{-1}y = B^{-1}Bx = x.$$

So

$$x = C^{-1}y = B^{-1}A^{-1}y,$$

that is,

$$C^{-1} = B^{-1}A^{-1}.$$

A.3.7 Determinant

There are several reasonable scalar measures of the "size" of a matrix. One is the *determinant*, denoted by det M or $|M|$.

The reader is referred to books on linear algebra for the full definition of the determinant.

In two-dimensional space, the area of a parallelogram bounded by vectors v and w with $v = (a, b)$ and $w = (c, d)$ is $ad - bc$, the determinant of the corresponding two-by-two matrix. Similar results hold for volumes and hypervolumes in higher dimensions.

A.4 Vector Differentiation

Let $x = (x_1\ x_2\ \ldots\ x_n)'$. Let f be a scalar function of x, $f(x)$.

The vector of partial derivatives of f with respect to $x_1, x_2, \ldots, x_n$ is denoted by

$$\text{grad } f = \frac{\partial f}{\partial x} = \left(\frac{\partial f}{\partial x_1}\ \frac{\partial f}{\partial x_2}\ \cdots\ \frac{\partial f}{\partial x_n}\right)'.$$

Sometimes this notation is written in-line, as, for example $\partial f/\partial x_i$. Analogous to $d\,ax\,/\,d\,x = a$, we have, for

$$a'x = a_1\,x_1 + a_2\,x_2 + \cdots + a_n\,x_n,$$

the rule

$$\partial a'x\,/\partial x = a = (a_1\ a_2\ \ldots\ a_n)'.$$

A.5 Paths

Let $x = (x_1\ x_2\ \ldots\ x_n)'$ and $x = p(t)$, where the variable t is time in a number of applications. The function $x(t)$ is called a *path* or *trajectory*. In terms of elements, $p(t) = (p_1(t)\ p_2(t)\ \ldots\ p_n(t))'$. The derivatives dp_i/dt of the elements p_i of p are in the vector $\partial p/dt = (dp_1/dt\ dp_2/dt\ \ldots\ dp_n/dt)'$. The gradient of $f(x)$ is $\text{grad } f(x) = (\partial f/\partial x_1\ \partial f/\partial x_2\ \ldots\ \partial f/\partial x_n)'$. In terms of $x = p(t)$, the function f is a composite function $f(p(t))$. The derivative of f with respect to t is

$$\frac{df}{dt} = \sum_{i=1}^{n} \frac{\partial f}{\partial x_i}\frac{dx_i}{dt} = \sum_{i=1}^{n} \frac{\partial f}{\partial p_i}\frac{dp_i}{dt} = \text{grad} \cdot \frac{\partial p}{dt},$$

where $\partial p/dt = (dp_1/dt\ dp_2/dt\ \ldots\ dp_n/dt)'$. This is a chain rule for vector functions.

These notions will be used in Appendix C on Lagrange multipliers.

A.6   Quadratic Forms

Given variables $v_1, v_2, \ldots, v_p$, a *quadratic form* in those variables is a linear combination of quadratic terms, that is, products of the form $v_i v_j$ and v_i^2. It is a function of the form

$$f(v_1, v_2, \ldots, v_p) = a_{11}v_1^2 + a_{12}v_1v_2 + \cdots + a_{p-1,p}v_{p-1}v_p + a_{pp}v_p^2.$$

A quadratic form can be written in terms of its vector $\boldsymbol{v}$ and its matrix $\boldsymbol{A}$ as $\boldsymbol{v}'\boldsymbol{A}\boldsymbol{v}$.

Remark. The matrix $\boldsymbol{A}$ of a quadratic form can always be taken as symmetric. For, if $\boldsymbol{A}$ is not symmetric, combine $a_{ij}v_iv_j + a_{ji}v_jv_i$ into $2a_{ij}^*v_iv_j$, where

$$a_{ij}^* = (a_{ij} + a_{ji})/2 \text{ and so } a_{ji}^* = a_{ij}^*.$$

Analogous to $d\,ax^2/dx = 2\,ax$, the derivative of a quadratic form is $\partial \boldsymbol{x}'\boldsymbol{A}\boldsymbol{x}/\partial \boldsymbol{x} = 2\boldsymbol{A}\boldsymbol{x}$.

A.7   Eigensystem

Given a $p \times p$ symmetric matrix $\boldsymbol{S}$, consider maximizing the quadratic form $\boldsymbol{w}'\boldsymbol{S}\boldsymbol{w}$ with respect to $\boldsymbol{w}$. There must be a condition on $\boldsymbol{w}$; otherwise taking one of its elements to be infinite would lead to a trivial maximum. Consider a condition of the form $\boldsymbol{w}'\boldsymbol{w} = c$. (This is the condition that the squared length of $\boldsymbol{w}$ be equal to c, that is, the length of $\boldsymbol{w}$ is $\sqrt{c}$.) Define a Lagrangian function for this problem (see Appendix C on Lagrange multipliers for details on this method) : $L(\boldsymbol{w}, \lambda) = \boldsymbol{w}'\boldsymbol{S}\boldsymbol{w} + \lambda(c - \boldsymbol{w}'\boldsymbol{w})$. The condition

$$\frac{\partial L}{\partial \boldsymbol{w}} = 2\boldsymbol{S}\boldsymbol{w} - 2\lambda\boldsymbol{w} = \boldsymbol{0}$$

leads to equations $\boldsymbol{S}\boldsymbol{w} = \lambda\boldsymbol{w}$, or

$$(\boldsymbol{S} - \lambda\boldsymbol{I})\boldsymbol{w} = \boldsymbol{0}.$$

This set of homogeneous simultaneous linear equations has a nontrivial solution (solution other than the zero vector) if and only if its coefficient matrix

$S - \lambda I$ has determinant equal to 0. This determinant is a polynomial of degree p (the order of S) in the variable λ, so setting it equal to zero gives an equation with p roots, the *eigenvalues* of S. Substituting the largest eigenvalue into

$$Sw = \lambda w$$

and solving gives w_1, the choice of w that maximizes the quadratic form. Note that if w is a solution, then so is dw, for any constant d. To make the solution unique, the side condition is needed. Pre-multiplying by w gives

$$w'Sw = \lambda w'w.$$

Taking the constant c in the side condition $w'w = c$ to be 1, that is, taking the side condition to be $w'w = 1$, gives $w'Sw = \lambda$, so that λ is then the value of the quadratic form.

The pair (λ, w) is called an *eigenpair*. The values λ are called *eigenvalues*. The eigenvalues of a symmetric matrix S are real and non-negative. Conventionally, these are numbered as

$$\lambda_1 \geq \lambda_2 \geq \ldots \geq \lambda_p \geq 0.$$

The corresponding eigenvectors are numbered

$$w_1, w_2, \ldots, w_p.$$

The set $\{(\lambda_v, w_v), v = 1, 2, \ldots, p\}$ of pairs (λ_v, w_v) is called the *eigensystem* of the given matrix. Eigenvalues and vectors are called also *characteristic* values and vectors, and *latent* values and vectors.

Several measures of the size of a matrix come to mind, including the determinant, the trace (sum of the diagonal elements), and the maximum eigenvalue. Note that these are all functions of the set of eigenvalues. The trace turns out to be their sum; the determinant, their product.

A.8 Transformation to Uncorrelated Variables

A.8.1 Covariance Matrix of a Linear Transformation of a Random Vector

Given a p-dimensional random vector (r.vec.) Y, let $X = MY$, where M is a $p \times p$ matrix. Then the covariance matrix of X, say Σ_X, is given by $\Sigma_X = M \Sigma_Y M'$, where Σ_Y is the covariance matrix of Y. To see this, write $C[X] = \mathcal{E}[(X - \mu_x)(X - \mu_x)'] = \mathcal{E}[(MY - M\mu_y)(MY - M\mu_y)'] = M \mathcal{E}[(Y - \mu_y)(Y - \mu_y)] M' = M C[Y] M'$.

A.8.2   Transformation to Uncorrelated Variables

A given r.vec. Y of p elements can be linearly transformed to a r.vec. U of p uncorrelated elements. (We say "a" r.vec. rather than "the" r.vec. because there are many ways to do this.) To see this, note that the eigensystem of Σ_Y, say

$$\{ (\lambda_v, \boldsymbol{w}_v), \; v = 1, 2, \ldots, p \},$$

can be expressed in matrix notation as

$$\boldsymbol{W}'\Sigma_Y \, \boldsymbol{W} = \Lambda = \mathrm{diag}(\lambda_1, \lambda_2, \ldots \lambda_p),$$

the diagonal matrix with diagonal elements λ_v, $v = 1, 2, \ldots, p$. The vector $U = \boldsymbol{W}'\, Y$ has covariance matrix $\Sigma_U = \boldsymbol{W}' \Sigma_Y \, \boldsymbol{W} = \Lambda$, which is diagonal. The off-diagonal elements, the covariances, are zero. Thus, the variables in the r.vec. U are uncorrelated.

A.8.3   Transformation to Uncorrelated Variables with Variances Equal to One

The p elements of a r.vec. Y are *linearly dependent* if the is a vector $\boldsymbol{a}$ such that $\boldsymbol{a}'\, Y = c$, a constant (with probability one). In this case, $\mathcal{E}[\boldsymbol{a}'Y] = c$. But $\mathcal{E}[\boldsymbol{a}'Y] = \boldsymbol{a}'\mu_y$. So $\mathcal{E}[\boldsymbol{a}'Y] = \boldsymbol{a}'\mu_y$, that is, $\mathcal{E}[\boldsymbol{a}'(y - \mu_y)] = 0$. The p elements of Y are *linearly independent* if there is no non-zero vector $\boldsymbol{a}$ such that $\mathcal{E}[\boldsymbol{a}'(Y - \mu_y)] = 0$.

When the p variables in Y are linearly independent, its covariance matrix Σ_y is non-singular and so its eigenvalues are positive. Then the r.vec. $U = \boldsymbol{W}'\, Y$ of uncorrelated elements can be further transformed to a vector Z whose elements are not only uncorrelated but also have variances equal to 1. To do this, let $Z = \Lambda^{-1/2}U$, where $\Lambda^{1/2} = \mathrm{diag}(\sqrt{\lambda_1}, \sqrt{\lambda_2}, \ldots, \sqrt{\lambda_p})$. Then $\Sigma_Z = \Lambda^{-1/2}\Lambda\Lambda^{-1/2} = I$. Note that $U = \boldsymbol{W}'\, Y$ and $Z = \Lambda^{-1/2}U = \Lambda^{-1/2}\boldsymbol{W}'Y$, so the transformation from Y to Z can be written in terms of a single matrix. as $Z = M\, Y$, where $M = \Lambda^{-1/2}\boldsymbol{W}$.

Further, given any orthogonal matrix P (that is, a matrix P such that $P'P = I = PP'$), take $F = PM$. Then FY has covariance matrix equal to $F\Sigma_Y F' = PM\Sigma_Y M'P' = PIP' = PP' = I$. Any transformation of the form $Z = P\Lambda^{-1/2}\boldsymbol{W}\, Y$ has identity covariance matrix. Note that then $Y = \boldsymbol{W}'\Lambda^{1/2}P'\, Z$, and one particular such transformation is simply $Y = \boldsymbol{W}'\Lambda^{1/2}Z$.

The eigensystem of Σ_y gives $\boldsymbol{W}'\Sigma_y \, \boldsymbol{W} = \Lambda$, that is,

$$\Lambda^{-1/2}\, \boldsymbol{W}'\Sigma_y \, \boldsymbol{W}\Lambda^{-1/2} = I.$$

The matrix $\boldsymbol{W}$ is orthogonal: $\boldsymbol{W}\boldsymbol{W}' = I = \boldsymbol{W}'\boldsymbol{W}$. Thus, upon pre-multiplying by $\boldsymbol{W}$ and post-multiplying by $\boldsymbol{W}'$, this gives $\boldsymbol{W}\boldsymbol{W}'\Sigma_y\boldsymbol{W}\boldsymbol{W}' = \boldsymbol{W}\Lambda\boldsymbol{W}'$, or $\Sigma_y = \boldsymbol{W}\Lambda\boldsymbol{W}'$, expressing Σ_y in terms of its eigensystem. Note that this can be written $\Sigma_y = \sum_{v=1}^{p} \lambda_v \boldsymbol{w}_v \, \boldsymbol{w}_v'$. This is called the *spectral representation* or *spectral decomposition* of the given covariance matrix Σ_y. Note that the representation is $\Sigma_y = C\,C'$, where $C = W\Lambda^{-1/2}$. Also, $\Sigma_y^{-1} = (C\,C')^{-1} = C'^{-1}C^{-1} = (C^{-1})'C^{-1}$.

A.9 Statistical Distance

Suppose that it is taken as axiomatic that Euclidean distance is appropriate for variables that are uncorrelated and have equal variances. This implies a distance for other variables. To see this, note first that the Euclidean distance between z and z_0 is taken as the length of the vector $d = z - z_0$. The length is $\sqrt{d'd}$; the squared length is $d'd$. Now, given the r.vec. Y, with covariance matrix Σ_y, it has been seen that there exists a matrix C (non-singular) such that the r.vec. $Z = CY$ has identity covariance matrix. Any such matrix C satisfies $C\,C' = \Sigma_Y^{-1}$. Thus, with $z_0 = C\,y_0$, the squared distance is

$$
\begin{aligned}
d'\,d &= (Z - z_0)'\,(Z - z_0) \\
&= (CY - C y_0)'\,(C Y - C\,y_0) \\
&= (Y - y_0)'\,C'\,C\,(Y - y_0) \\
&= (Y - y_0)'\,\Sigma_Y^{-1}\,(Y - y_0) \\
&= D^2(Y, y_0; \Sigma_Y),
\end{aligned}
$$

where, given vectors u and v and a non-singular matrix M, the function $D^2(u, v; M)$ is $(u - v)'\,M^{-1}\,(u - v)$. The squared distance D^2 is squared *statistical distance*, or *Mahalanobis distance* (Mahalanobis 1936). It is distance that is adjusted for correlations and unequal variances.

A.10 Appendix Exercises

A.14 What is the cosine of the angle between the vectors $(1\ \ 1)$ and $(1\ -1)$?

A.15 Are the vectors $(1\ 1\ 1)$ and $(1\ -1\ 0)$ orthogonal?

A.16 (continuation) Show that the vector $(1\ 1\ -2)$ is orthogonal to these two.

A.17 Show that $(1 + v^2)^{-1} = 1 - \alpha v^2$, where $\alpha = 1/(1 + v^2)$.

A.18 Rank-one perturbation. Show that the inverse of $I + vv'$ is $I - \alpha vv'$, where $\alpha = 1/(1 + v'v)$.

A.19 (continuation) Generalize this to $M + vv'$, where M is non-singular.

A.20 (continuation) By taking M to be of the form $M = XX'$ and v to be the vector of values of a new explanatory variable to be included, show how this can be applied to stepwise regression.

A.21 What is the inverse of $I + uv'$?

A.22 Expand $[I - vv']^{-1}$ in powers of $v'v$ when $v'v < 1$.

A.23 Show that $c\,v$, where c is a constant, satisfies the eigenpair equation for $I + vv'$. What is the corresponding eigenvalue?

A.24 (continuation) Verify that $c\,(v_2 \quad -v_1 \quad 0 \ \ldots \ 0)'$, where c is a constant, is an eigenvector of $I + vv'$ and that 1 is the corresponding eigenvalue.

A.25 (continuation) Verify that $c(v_3 \quad 0 \quad -v_1 \quad 0 \ \ldots \ 0)'$ is an eigenvector of $I + vv'$ and that 1 is the corresponding eigenvalue.

A.11 Bibliography

Anderson, T. W. (2003) *An Introduction to Multivariate Statistical Analysis. 3rd ed.* John Wiley & Sons, New York. (First edition, 1958.)

Hohn, Franz E. (2003). *Elementary Matrix Algebra. 3rd ed.* Dover Publications, Mineola, New York (1st ed. 1958, 2nd ed., 1973, Macmillan, New York).

Lay, David C. (2012). *Linear Algebra and Its Applications. 4th ed.* Addison-Wesley, Boston, MA.

Mahalanobis, Prasanta Chandra (1936). On the generalised distance in statistics. *Proceedings of the National Institute of Sciences of India,* **2,** 49–55.

Marcus, Marvin, and Minc, Henryk (1965). *Introduction to Linear Algebra.* Macmillan, New York. Republication 1988, Dover Publications, New York.

A.12 Further Reading

For further study of vectors and matrices, the reader may wish to consult text-books such as those by Hohn(2003) or Lay (2012). Appendix A in Anderson (2003) is concise and helpful.

B

Normal Distributions

CONTENTS

This appendix relates to univariate and multivariate Normal distributions. Vectors and matrices, as discussed in Appendix A, are used here.

B.1   Some Results for Univariate Normal Distributions

B.1.1   Definitions

The family of Normal distributions is a two-parameter family. The parameters are the mean μ and the variance σ^2. The standard deviation is the square root of the variance. The *standard Normal distribution* is the Normal distribution with mean 0 and variance 1. Its p.d.f. is

$$\phi(z) = \frac{1}{\sqrt{2\pi}} e^{-z^2/2}, \quad -\infty < z < \infty.$$

The c.d.f. is $\Phi(z) = \int_{-\infty}^{z} \phi(z)\,dz = \Pr\{Z \leq z\}$, where Z is a random variable having the standard Normal distribution.

If a random variable Y has a Normal distribution with mean μ and variance σ^2, then sometimes for brevity this is denoted by $Y \sim \mathcal{N}(\mu, \sigma^2)$. If $Y \sim \mathcal{N}(\mu\sigma^2)$, then Y is distributed as $\mu + \sigma Z$, where Z is distributed according to the standard Normal distribution, that is, $Z \sim \mathcal{N}(0, 1)$. The p.d.f. of Y is

$$\phi(y; \mu, \sigma^2) = \frac{1}{\sigma\sqrt{2\pi}} e^{-(y-\mu)^2/2\sigma^2}, \quad -\infty < y < \infty.$$

The c.d.f. is denoted by $\Phi(y; \mu, \sigma^2)$.

B.1.2   Conditional Expectation

The conditional expectation

$$\begin{aligned}
\mathcal{E}[Z \mid a < Z < b] &= \int_a^b z\phi(z)\,dz \\
&= \int_a^b z \frac{1}{\sqrt{2\pi}} e^{-z^2/2}\,dz \,/\, [\Phi(b) - \Phi(a)] \\
&= \int u \frac{1}{\sqrt{2\pi}} e^{-u}\,du \,/\, [\Phi(b) - \Phi(a)] \\
&= \frac{1}{\sqrt{2\pi}} [-e^{-u}] \,/\, [\Phi(b) - \Phi(a)] \\
&= \frac{1}{\sqrt{2\pi}} [-e^{-z^2/2}]_a^b \,/\, [\Phi(b) - \Phi(a)] \\
&= \frac{\phi(a) - \phi(b)}{\Phi(b) - \Phi(a)},
\end{aligned}$$

where the integration is by substitution with $u = z^2/2$, $du = z\,dz$. If $b = \infty$, then the result gives

$$\mathcal{E}[Z \mid Z > a] = \phi(a)/[1 - \Phi(a)].$$

If $a = -\infty$, then the result gives

$$\mathcal{E}[Z \mid Z < b] = -\phi(b)/\Phi(b).$$

If Y is distributed according to the Normal distribution with mean μ and variance σ^2, the conditional expectation of Y, given that $y_0 < Y, y_1$, is

$$
\begin{aligned}
\mathcal{E}[Y \mid y_0 < Y < y_1] &= \mathcal{E}[\mu + \sigma Z \mid y_0 < \mu + \sigma Z < y_1] \\
&= \mathcal{E}[\mu + \sigma Z \mid (y_0 - \mu)/\sigma < Z < (y_1 - \mu)/\sigma] \\
&= \mu + \sigma \mathcal{E}[Z \mid (y_0 - \mu)/\sigma < Z < (y_1 - \mu)/\sigma] \\
&= \mu + \sigma \frac{\phi(z_0) - \phi(z_1)}{\Phi(z_1) - \Phi(z_0)},
\end{aligned}
$$

where $z_0 = (y_0 - \mu)/\sigma$ and $z_1 = (y_1 - \mu)/\sigma$. If $y_1 = \infty$, then

$$
\mathcal{E}[Y \mid Y > y_0] = \mu + \sigma \frac{\phi(z_0)}{1 - \Phi(z_0)},
$$

where $z_0 = (y_0 - \mu)/\sigma$. If $y_0 = -\infty$, then

$$
\mathcal{E}[Y \mid Y < y_1] = \mu - \sigma \frac{\phi(z_1)}{\Phi(z_1)},
$$

where $z_1 = (y_1 - \mu)/\sigma$.

B.1.3 Tail Probability Approximation

For large z, the tail probability $\Pr\{Z > z\} \approx \phi(z)/z$. This is a sharp upper bound (see Feller 1957, page 166). For example, for $z = 3$, this gives .001477, whereas the exact answer is about .00135.

B.2 Family of Multivariate Normal Distributions

The family of multivariate Normal distributions plays a central role in much of the field of statistics.

A particular member of the family of multivariate Normal distributions is specified by its parameters, the *mean vector* and *covariance matrix*. The *mean vector* $\boldsymbol{\mu}$ is the vector of means of the p variables, μ_v, $v = 1, 2, \ldots, p$, and the covariance matrix $\boldsymbol{\Sigma}$ is the matrix of covariances σ_{uv}, $u, v = 1, 2, \ldots, p$. The parameter σ_{vv} is the variance of the v-th variable; it is the square of the standard deviation σ_v; that is, $\sigma_{vv} = \sigma_v^2$.

Because each covariance is the product of the corresponding correlation, times the product of the standard deviations, that is, $\sigma_{uv} = \rho_{uv}\sigma_u\sigma_v$, specifying the covariance matrix is equivalent to specifying the correlations and standard deviations.

An interesting and important characterizing property of the family of Normal distributions is that if a random vector $\boldsymbol{Y}$ has a multivariate Normal distribution, then every linear combination of the elements of $\boldsymbol{Y}$ has a univariate Normal distribution.

Another important property is linearity of the conditional expectation. If $(X_1, X_2, \ldots, X_p, Y)$ have a multivariate Normal distribution, then $\mathcal{E}[Y \mid x_1, x_2, \ldots, x_p]$ is of the form $\alpha + \beta_1 x_1 + \beta_2 x_2 + \cdots + \beta_p x_p$.

B.3 Role of D-Square

The multivariate Normal probability density function (p.d.f.) involves Mahalanobis D-squared, $D^2(x, \mu; \Sigma)$, which is the quadratic form

$$D^2(x; \mu, \Sigma) = (x - \mu)' \Sigma^{-1}(x - \mu),$$

where x is the vector of values of the variables, μ is the mean vector, and Σ is the covariance matrix. In scalar notation,

$$D^2 = \sum_{u=1}^{p} \sum_{v=1}^{p} \sigma^{uv} (x_u - \mu_u)(x_v - \mu_v),$$

where σ^{uv} is the (u, v)-th element of Σ^{-1}.

Denote this quadratic form by Q for short. Then the p.d.f. $\varphi(x; \mu, \Sigma)$ is of the form

$$\varphi = \text{Const.} \, e^{-Q/2},$$

where

$$\text{Const.} = [\, (2\pi)^{-p/2} (\det \Sigma)^{-1/2} \,],$$

"det" denotes determinant, and p is the number of variables. When $p = 1$, $\text{Const.} = 1/\sqrt{2\pi\sigma^2} = 1/\sigma\sqrt{2\pi}$.

The quantity Q is the square of the statistical (Mahalanobis) distance between x and μ. The larger this distance, the smaller the probability density; the density decreases exponentially with the square of the distance. The Normal density is denoted by φ and is

$$\begin{aligned}
\varphi(x; \mu, \Sigma) &= \text{Const.} \, e^{-Q/2} \\
&= (2\pi)^{-p/2} (\det \Sigma)^{-1/2} \exp[(-1/2)D^2(x; \mu, \Sigma)].
\end{aligned}$$

In the univariate case, the quadratic form simplifies to $Q = z^2$, where $z = (x - \mu)/\sigma$, the so-called z score or *standard score*.

B.4 Bivariate Normal Distributions

A *bivariate* Normal distribution (joint Normal distribution of $p = 2$ variables) is specified by giving the values of five parameters, namely,

- the two means,

- the two standard deviations, and

- the correlation (or, equivalently, the covariance).

Example: Height and weight

Suppose for a population of adult males, height H and weight W are jointly Normally distributed with $\mu_H = 68$ in., $\sigma_H = 2.5$ in., $\mu_W = 165$ lbs., $\sigma_W = 25$ lbs., and $\rho_{H,W} = +.4$. Then, letting $\mathcal{C}$ denote covariance, $\mathcal{C}[H, W]$ is $\sigma_{HW} = \rho_{HW}/\sigma_H \sigma_W = (+.4)(2.5)(25) = +25.0$. (The units of the covariance are then lb. $\times$ in.)

B.4.1   Shape of the p.d.f.

Because the density decreases exponentially with the square of the distance between (x, y) and the mean vector, the bivariate Normal p.d.f. gives a surface $z = f(x, y)$ that is bell-shaped. The "bell-shaped" univariate Normal distribution is a cross-section of the bivariate Normal bell.

B.4.2   Conditional Distribution of Y Given X

Another way to specify a bivariate normal distribution is in terms of the conditional distribution of Y given X and the marginal distribution of X. Analogous to $\Pr(A \cap B) = \Pr(A)\Pr(B|A)$, for p.d.f.s we have $f(x, y) = f(x)f(y|x)$. It is often easy to describe these two and then multiply them to obtain the joint p.d.f.

B.4.3   Regression Function

The conditional distribution of Y given X involves the mean of Y when $X = x$; as we know, this function is called the "regression function." When X and Y have a joint Normal distribution, the regression function is $\mathcal{E}[Y \mid x] = \alpha + \beta x$, where $\beta = \sigma_{xy}/\sigma_x^2$ and $\alpha = \mu_y - \beta \mu_x$.

The variance $\sigma_{y|x}^2$ of the conditional distribution is a constant (not varying with x) and is equal to $\sigma_{y|x}^2 = \sigma_y^2(1-\rho_{xy}^2) = \sigma_y^2 - \sigma_{xy}^2/\sigma_x^2 = \sigma_y^2 - \beta^2 \sigma_x^2$. The parameter $\sigma_{y|x}$ is called the *standard error of regression* and is denoted also by $\sigma_{y \cdot x}$.

Example: Height and weight, continued

In the example, the mean weight for men of height h is $\alpha + \beta h$, where

$$\beta = \sigma_{HW}/\sigma_H^2 = +25.0/6.25 = 4.0 \text{ lbs. per in.}$$

$$\alpha = \mu_W - \beta\mu_H = 165 - (4)(68) = 165 - 272 = -107 \text{ lbs.;}$$

that is, the mean weight for men of height h is $4h - 107$ lbs. For example, if $h = 70$ in., this is

$$4(70) - 107 = 280 - 107 = 173 \text{ lbs.}$$

The conditional variance $\sigma_{wt|ht}^2 = 25^2 - 4.0^2(6.25) = 625 - 100 = 525; \sigma_{wt|ht} = \sqrt{525}$, or about 22.9 lbs.

B.5   Other Multivariate Distributions

There are many continuous multivariate distributions in addition to the Normal. It is interesting to construct one from elementary considerations. Let the conditional distribution of Y given $X = x$ be exponential with parameter x:

$$f(y|x) = x \exp(-xy), \, y > 0, \, x > 0.$$

Let X have an exponential distribution with parameter λ :

$$f(x) = k \exp(-kx), \, x > 0.$$

Then,

$$
\begin{aligned}
f(x,y) &= f(x)f(y|x) \\
&= k \exp(-kx)x \exp(-xy) \\
&= k\,x \exp[-(kx + xy)] \\
&= k\,x \exp[-x(y + k)], x > 0, y > 0.
\end{aligned}
$$

B.6   Summary

B.6.1 Concepts

This appendix concerns the family of multivariate Normal distributions.

1. The *parameters* of a multivariate Normal distribution are the mean vector and the covariance matrix.

2. The bivariate Normal distribution has five parameters: two means, two variances, and a covariance. Specifying the variances and covariance is equivalent to specifying the correlation and the two standard deviations.

B.6.2 Mathematics

1. The multivariate Normal p.d.f. depends upon x only through Mahalanobis D-square.

2. When X and Y have a bivariate Normal distribution, the regression function of Y on X is a linear function of x.

3. If $(X_1, X_2, \ldots, X_p, Y)$ have a multivariate Normal distribution, then $\mathcal{E}[Y \mid x_1, x_2, \ldots, x_p]$ is of the form $\alpha + \beta_1 x_1 + \beta_2 x_2 + \cdots + \beta_p x_p$.

B.7 Appendix B Exercises

B.7.1 Applied Exercises

B.1 The *partial correlation coefficient* between X and Y, taking account of T, can be written as

$$\rho_{xy.t} = (\rho_{xy} - \rho_{xt}\rho_{ty})/\sqrt{1 - \rho_{xt}^2}\sqrt{1 - r_{ty}^2}.$$

If $\rho_{xy} = .8, \rho_{xt} = .6$, and $\rho_{ty} = .8$, compute $\rho_{xy.t}$.

B.2 (continuation) If $\rho_{xy} = 0$, $\rho_{xt} = .6, \rho_{ty} = .8$, compute $\rho_{xy.t}$.

B.3 Download the Fisher iris data (Anderson 1935, Fisher 1936) from the University of California–Irvine (UC - I) dataset repository at the URL http://archive.ics.uci.edu/ml/datasets.html. The dataset consists of 150 observations, 50 observations on each of three species of iris. The $p = 4$ variables are petal and sepal length and width. Use software to estimate the mean vectors and covariance matrices in the three species.

B.4 (continuation) Do the elements of the three mean vectors look different? For all four variables?

B.5 (continuation) Do the covariance matrices seem similar across the three species?

B.7.2   Mathematical Exercises

B.6 In scalar notation, write Mahalanobis D-squared for the bivariate case.

B.7 (continuation) Write the p.d.f. for the bivariate case.

B.8 If for adult males the systolic blood pressure has a mean of 120 and a standard deviation of 17 and the diastolic blood pressure has a mean of 80 and a standard deviation of 11 and the correlation is .8, find the regression function of systolic on diastolic and the conditional variance.

B.9 If for adult males the systolic blood pressure has a mean of 120 and a standard deviation of 17 and the age has a mean of 44 and a standard deviation of 11 and the correlation is .35, find the regression function of systolic on age and the conditional variance. The regression functionf $\mathcal{E}[\text{Sys} \,|\, \text{Age}]$ is of the form $\alpha + \beta$ Age. Is the estimated regression function close to $100 + \text{Age}/2$, corresponding to the simple rule, 100 plus half the age?

B.10 In the bivariate exponential example, find the p.d.f. of $X \,|\, Y$, that is, the conditional p.d.f. of X given Y. *Hints:* It is $f(x|y) = f(x,y)/f(y)$. First find $f(y)$ by integrating $f(x,y)$ with respect to x.

B.8   Bibliography

Anderson, Edgar (1935). The irises of the Gaspé peninsula. *Bulletin of the American Iris Society,* **59,** 2–5.

Anderson, T. W. (1958). *An Introduction to Multivariate Statistical Analysis.* John Wiley & Sons, New York. (3rd edition, 2003).

Feller, William (1957). *An Introduction to Probability and Its Applications. Vol. 1. 2nd ed.* John Wiley & Sons, New York. (1st ed. 1950, 3rd edition 1968.)

Fisher, R.A. (1936). The use of multiple measurements in taxonomic problems. *Annals of Eugenics,* **7,** 179–188.

Johnson, Richard A., and Wichern, Dean W. (2007). *Applied Multivariate Statistical Analysis. 6th ed.* Prentice Hall (Pearson), Upper Saddle River, NJ.

B.9 Further Reading

Books on multivariate statistical analysis contain thorough discussions of the family of multivariate Normal distributions. There are books at different levels. The first, definitive, text is that by T. W. Anderson (1958).

C

Lagrange Multipliers

CONTENTS

C.1   Notation

(See also Appendix A on Vectors and Matrices.)

The symbol x denotes an n-dimensional vector, with elements x_1, x_2, ..., x_n. The functions f and g are scalar functions of x. The *dot product* (*inner product, scalar product*) of vectors u and v is denoted by $u \cdot v$. (See Appendix A - Vectors and Matrices.)

The null vector is denoted by 0. The *gradient* of $f(x)$, denoted by $\mathbf{grad}\, f(x)$, is the vector of partial derivatives of f with respect to the elements of x :

$$\mathbf{grad}\, f = (\partial f/\partial x_1 \quad \partial f/\partial x_2 \quad \cdots \quad \partial f/\partial x_n).$$

C.2   Optimization Problem

The following general optimization problem is considered.

Maximize $f(x)$, subject to x in S, where $S = \{x : g(x) = 0\}$.

Remarks. (i) This is called "maximizing f subject to the side condition (constraint) $g(x) = 0$." (ii) The same mathematics applies to the problem of minimizing f subject to a constraint. The point is that a stationary point of a function, the Lagrangian, incorporating the constraint, is found.

Lemma. Suppose f attains its maximum on S at a point x^* not on the boundary of S. Then

$$\mathbf{grad}\, f(x^*) \;=\; k\, \mathbf{grad}(x^*)$$

for some constant k.

Theorem. A point x^* where $f(x)$ has its maximum value on the surface $g(x) = 0$ satisfies $g(x^*) = 0$ and $\mathbf{grad}\,(f - kg)(x^*) = 0$ for some constant k.

Remark. The application of the theorem is in maximizing $f(x)$ subject to the *constraint* $g(x) = 0$. One forms the function

$$f(x) - kg(x),$$

takes its partial derivatives with respect to the elements of x, and sets them equal to zero. One then solves the resulting equations, together with the equation $g(x) = 0$. The constant k is called a *Lagrange multiplier*.

Proof of Lemma. Let $x = p(t)$ be any path lying in S and passing through x^*, that is, $x^* = p(t^*)$ for some t^*. Then $f[p(t)]$ has its maximum at t^* and its derivative must be zero there. The chain rule for vector functions gives

$$df/dt \;=\; \mathbf{grad}\, f \cdot \partial p/dt.$$

Because $df/dt = 0$ at x^*, then $\mathrm{grad}\, f(x^*) \cdot \partial p/dt|_{t=t*} = 0$, so $\mathrm{grad}\, f(x^*)$ is orthogonal to $\partial p/dt|_{t=t*}$, that is, $\mathrm{grad}\, f(x^*)$ is perpendicular to the path at x^*. Therefore, $\mathrm{grad}\, f(x^*)$ lies in the direction normal to S at x^*. But $\mathrm{grad}\, g(x^*)$ also is normal to S at x^*. Therefore, $\mathrm{grad}\, f(x^*)$ and $\mathrm{grad}\, g(x^*)$ are parallel, that is, there exists a constant k such that $\mathrm{grad}\, f(x^*) = k\, \mathrm{grad}\, g(x^*)$.

Proof of Theorem. We have $\mathrm{grad} f = k\, \mathrm{grad}\, g$ iff. $\mathrm{grad} f = k\, \mathrm{grad}\, g = 0$ iff. $\mathrm{grad}\,(f - kg) = 0$.

C.3   Bibliography

Loomis, Lynn (1977). *Calculus. 2nd ed.* Addison-Wesley, Reading, MA. (1st ed. 1974).

C.4 Further Reading

The calculus book by Loomis is especially highly recommended in general. In particular, see pages 590–591 on Lagrange multipliers (or page 696, Exercise 21, in the first edition).

D

Abbreviations and Symbols

CONTENTS

D.1   Abbreviations

D.1.1   Statistics

c.d.f.	Cumulative distribution function
d.f.	Degrees of freedom
EM	Expectation-Maximization (algorithm)
HMM	Hidden Markov model
i.i.d.	Independent and identically distributed (random variables)
m.e.	Margin of error
p.d.f.	Probability density function
p.m.f.	Probability mass function
r.v.	Random variable
r.vec.	Random vector
SD	Standard deviation
SE	Standard error

D.1.2   General

AAAS	American Association for the Advancement of Science
ASA	American Statistical Association
IEEE	Institute of Electrical and Electronics Engineers
iff.	if and only if

D.1.3 Finance

CAPM	Capital Asset Pricing Model
DJIA	Dow Jones Industrial Average
DREVX	Ticker symbol for Dreyfus Fund, Inc.
EFT	Exchange Traded Fund
GBM	Geometric Brownian motion model
MDY	Ticker symbol for Standard and Poor's midcap 400 ETF
ROR	Rate of return
S&P500	Standard & Poor's 500 Stock Composite Index
SPDR	Standard & Poor's Depositary Receipts
SPY	Ticker symbol for Standard and Poor's 500 ETF

D.2   Symbols

D.2.1   Statistics

Boldface lower-case such as $\boldsymbol{v}, \boldsymbol{w}, \boldsymbol{x}$ is used for vectors; upper-case letters such as $\boldsymbol{A}, \boldsymbol{S}$, for matrices.

X, Y, Z	r.v.s.
	It is often helpful to denote a r.v. by an upper-case letter such as Y, a realized value it by the corresponding lower-case letter y, and a sample of n values by $y_1, y_2, \ldots, y_n$.
$p_X(v_j)$	p.m.f. of the discrete r.v. X with values v_j, $j = 1, 2, \ldots, m$
$f_X(v)$	p.d.f. of the continuous r.v. X evaluated at v
$f_{X,Y}(x, y)$	Joint p.d.f. of the r.v.s X, Y evaluated at x, y
$f_{Y|X}(y|x)$	Conditional p.d.f. of the r.v. Y, given $X = x$, evaluated at y
μ_x or $\mathcal{E}[X]$	Expected value, mathematical expectation, mean
σ_x^2 or $\mathcal{V}[X]$	Variance
σ_x or $\mathrm{SD}[X]$	Standard deviation
$\mathcal{C}[X, Y]$:	Covariance. Denoted also by σ_{xy}.
$\mathrm{Corr}[X, Y]$	Correlation of the r.v.s X and Y. Denoted also by ρ_{xy}.
$\beta_{y \cdot x}$	Coefficient of regression of Y on x
$\rho_{yz \cdot x}$	Partial correlation coefficient of y and z, adjusted for x
$\bar{x}$	Sample mean
s_x	Sample standard deviation
s_x^2	Sample variance
s_{xy}	Sample covariance
$\hat{\rho}_{xy}$	Sample correlation
$\boldsymbol{Y}$	r.vec., with elements $Y_1, Y_2, \ldots, Y_p$
$\boldsymbol{\Sigma_y}$ or $\mathcal{C}[\boldsymbol{Y}]$	Covariance matrix of the r.vec. $\boldsymbol{Y}$

D.2.2   Finance

P_t	Price at time t
p_t	Logarithm (natural log) of price at time t
	(Note that this use of P and p breaks our convention
	about lower-case letters being realized values
	of the corresponding upper-case letter.)
R_t	ROR at time t
r_t	Continuous ROR at time t
	(Note that this use of R and r breaks our convention
	about lower-case letters being realized values
	of the corresponding upper-case letter.)
O_t	Opening price of a stock for time period t (on day t, say)
C_t	Closing price of a stock for time period t
H_t	Highest price of a stock for time period t
L_t	Lowest price of a stock for time period t

AAAS $+ u = \Delta$
(AAAS plus you equals change!)

—American Association for the Advancement of Science
T-shirt slogan

Index

Printed in the United States
by Baker & Taylor Publisher Services